NMR DATA PROCESSING

NMR DATA PROCESSING

JEFFREY C. HOCH
ALAN S. STERN
Rowland Institute for Science
Cambridge, Massachusetts

WILEY-LISS

A JOHN WILEY & SONS, INC., PUBLICATION
New York • Chichester • Brisbane • Toronto • Singapore

Address All Inquiries to the Publisher
Wiley-Liss, Inc., 605 Third Avenue, New York, NY 10158-0012

Library of Congress Cataloging-in-Publication Data

Hoch, Jeffrey C.
   NMR data processing / Jeffrey C. Hoch and Alan S. Stern.
     p.   cm.
   ISBN 0-471-03900-4 (cloth : alk. paper)
   1. Nuclear magnetic resonance spectroscopy—Data processing.
I. Stern, Alan S., 1958–   . II. Title.
QD96.B8H63 1996
543'.0877—dc20   96-4981
                                                              CIP

CONTENTS

PREFACE

Nuclear magnetic resonance spectroscopy is one of the most powerful nondestructive techniques available today for probing the structure of matter. It appears in many guises. High-resolution NMR is valuable for determining the structure of chemical compounds; it is the only method capable of determining the three-dimensional structure of biomolecules in solution. Solid-state NMR is capable of providing precise structural information on polymeric, powdered, crystalline, and polycrystalline substances. Magnetic resonance imaging (MRI) is rapidly becoming one of the most important noninvasive diagnostic tools in medicine, and is also finding important uses for imaging nonbiological materials. NMR microscopy promises to bring similar imaging to the microscopic scale.

Aside from their common reliance on the fundamental physics of nuclear spins in an external magnetic field, the diverse applications of NMR share another feature. Without modern computers to acquire, process, and analyze the data from NMR experiments, the applications could never have developed. Richard Ernst, recipient of the 1991 Nobel Prize in Chemistry for his contributions to modern NMR spectroscopy, put it more succinctly: "Without Computers, No Modern NMR" [1].

This book is concerned with NMR data processing. The major task associated with processing NMR data is *spectrum analysis.* In modern NMR experiments, the nuclear spins in the sample are perturbed from equilibrium, and the time-dependent response of the entire collection of spins is measured through the voltage they induce in a detector, such as a coil. This free induction decay (FID) contains contributions from all of the spins in the sample, and accordingly is sometimes called an *interferogram.* Spectrum analysis involves resolving the "sound" produced by the chorus of spins to characterize, to the extent possible, contributions from individual members in terms of the frequencies, amplitudes, and phases of their time-varying responses.

Spectrum analysis plays an important role in all of the physical sciences, in engineering, in communication, and even in consumer electronics. It remains an active area of research, and new developments and algorithms will no doubt continue to improve the analysis of NMR data. The aim of this book is not to try to achieve a comprehensive treatment of this subject, for even if such a goal were within the authors' grasp, it would cease to be comprehensive with the first new development. Rather, the aim of this book is to provide an exposition of some fundamental principles and a set of tools to apply those fundamentals, together forming a framework for analysis of NMR data and a basis for the development of new methods.

The discrete Fourier transform (DFT) plays a central role in spectrum analysis and was crucial to the development of all modern applications of NMR. To emphasize the importance of the DFT, we devote two chapters to the subject: one covering fundamental aspects of the DFT and the other concerning its application to NMR. We refer to DFT-based methods as "classical" spectrum analysis. Two subsequent chapters are devoted to "modern" spectrum analysis: linear prediction and maximum entropy reconstruction, powerful methods that avoid some of the difficulties encountered with the DFT. A chapter devoted to other, less widely used, methods of modern spectrum analysis describes image-based methods, Bayesian techniques, and the discrete wavelet transform. The book concludes with a chapter on visualization, quantification, and error analysis—three topics that are distinct from spectrum analysis but indispensable in data processing.

A subtitle for this book could be "Inside the Black Box." Our aim is to remove some of the mystery behind the techniques embodied in modern NMR data processing software. Very few people will need, or be inclined, to implement any of the data processing techniques described in this book. Yet this book is not meant only for software developers, but also for users of NMR software. Armed with an understanding of the methods, a user is better equipped to exploit the full power of NMR software and to avoid the pitfalls of misapplication or misinterpretation.

This book does not expound on the principles of the NMR experiment itself, for which there are a number of excellent sources. Nor does it discuss the meaning or interpretation of the experimental data once the primary analysis is complete—such a vast topic would require a treatment as broad-ranging as the set of applications of NMR itself. It concentrates solely on that middle area so often taken for granted: the reduction of raw data to a form that is usable for further scientific investigation. Since the principle focus of our laboratory is the study of proteins by NMR, the treatment is noticeably and unavoidably biased. Nevertheless, the large majority of the methods described are applicable to all of the diverse forms of NMR spectroscopy.

Isaac Asimov was quoted as saying that the best way to learn a subject is to write a book on it. (Having written hundreds of books in his lifetime, by this measure Asimov was a very learned man.) Writing this book has indeed been a learning experience. While our main goal in writing this book is to convey what we know about NMR data processing, we realize that it will also convey our prejudices and expose what we don't know. It is our sincere hope that any shortcomings are small enough to leave the final result both useful and enjoyable.

There are many people who deserve acknowledgment for making this book possible. Whether through collaborations or informal discussions, lending advice, software, molecules, data, or other assistance, the support they provided made our task not only easier but also more enjoyable. To Mike Burns, Roger Chylla, Lou Cincotta, Rob Clubb, Peter Connolly, David Donoho, Jim Foley, Theo Greene, Tim Havel, Sven Hyberts, George Maalouf, Claudia Mastroianni, MaryAnn Nilsson, John Osterhout, Marcella Oslin, David Ruben, Bob Savoy, Jay Scarpetti, Peter Schmieder, Craig Schaefer, Sekhar Talluri, and Gerhard Wagner, we offer our most sincere thanks. We are grateful to Richard Ernst and Oleg Jardetzky for providing the photographs in Figures 1.1 and 7.11, respectively. We also gratefully acknowledge the support of the National Institutes of Health, which provided some of the computer equipment used to complete this book (Grant GM-47467, G. Wagner, P.I.). Throughout, the able assistance of the editorial staff at John Wiley & Sons has been invaluable.

The Rowland Institute for Science is our intellectual home. For the support, freedom, stimulation, and encouragement it provides, we consider ourselves to be uniquely fortunate.

Special thanks are due to Linda Turner Stern and Loren and Emma Lou Hoch. Monet found inspiration at Giverny; the Hoch garden at Akron is no less inspiring, and without it this book most certainly would not have been written.

JEFFREY C. HOCH
ALAN S. STERN

August 1995
Cambridge, Massachusetts

Le Jardin à Akron.

NMR DATA PROCESSING

1

INTRODUCTION

1.1. SOME HISTORY

The process of determining the frequency components contained in the free induction decay (FID) is an integral part of the NMR experiment. This has not always been so. Prior to 1965, NMR spectra were routinely determined by measuring the resonant absorption of radio frequency (RF) radiation, either at fixed frequency while varying the magnetic field (field-swept NMR), or at fixed magnetic field while varying the frequency (frequency-swept NMR). The importance of spectral analysis of time series data in NMR was cemented in 1965 by Richard Ernst and Weston Anderson, then at Varian Associates. They proposed that a more efficient way to determine the NMR spectrum is to measure the time-varying response of the nuclear spins following the application of an RF pulse sufficient to excite all of the spins in the sample simultaneously, then Fourier transform the result to obtain the spectrum.

Although the relationship between the discrete Fourier transform (DFT) of a time series and its spectrum had long been known to physicists, this was nevertheless an ambitious proposal at the time. The first experiments conducted by Ernst and Anderson required measuring the FID and transferring the data, by means of paper tape and punched cards, to a computer "service bureau" that was miles away from the NMR spectrometer—hardly an efficient process! Yet they firmly established the advantages that could be realized by this method, and they eventually managed to obtain a laboratory minicomputer that they interfaced to the spectrometer (Fig. 1.1). It was not long before the first commercial "FT-NMR" spectrometer appeared.

The year 1965 proved to be momentous for more than Ernst's pulsed NMR proposal: It also marked the publication of an algorithm, by James Cooley and

1

Fig. 1.1. Richard Ernst served as the spectrometer-computer interface for some of the first FT-NMR experiments.

John Tukey, for computing the DFT in order $N\log(N)$ computations, a remarkable improvement over the order N^2 operations implicit in the definition of the DFT (Chapter 2). This algorithm and subsequent modifications, now known simply as fast Fourier transforms (FFTs), was another enabling development in the evolution of modern NMR. Richard Ernst has mentioned that he did not become aware of the FFT algorithm until nearly a year after his initial experiments with FT-NMR, so his proposal must have seemed particularly audacious, and practical only for short data records. The FFT made practical the computation of the DFT for arbitrarily long data records on ordinary laboratory computers of the time.

It turns out that the FFT has a much richer history than was widely known in 1965. Some of the underlying ideas had been published as early as 1903, and other scientists were using algorithms similar to Cooley and Tukey's in 1965. Nevertheless, it was Cooley and Tukey's paper that was most responsible for the promulgation of FFT algorithms. The consequences have been profound, not just for NMR spectroscopy, but for all areas of science and engineering.

Development of the FFT algorithms would have had little significance were it not for digital computers to execute them. The first computer that Ernst and Anderson attached to their NMR spectrometer was a PDP-8, a computer with a 12-bit word length capable of computing a short FFT in seconds. Today typical laboratory workstations contain central processing units consisting of a single integrated circuit, with word lengths of 32 or 64 bits and capable of

Fig. 1.2. By 1981, the importance of computers in NMR was clear even to marketing types.

computing an FFT in milliseconds. This phenomenal improvement in computational speed has certainly made NMR spectroscopy more convenient, but more importantly it has made practical such developments as multidimensional NMR and NMR imaging, in which thousands or millions of DFTs are required to analyze the data. The importance of modern computers for NMR was not lost on the marketers at one computer company, who included it (albeit with tongue in cheek) in an advertisement that appeared in the popular magazine Newsweek in the early 1980s (Fig. 1.2).

1.2. SOME DEFINITIONS

1.2.1. Spectrum Analysis

We have used the term "spectrum analysis" a bit loosely up to now. In the broadest sense, *spectrum analysis* is the process of analyzing a time-varying signal to characterize the frequencies, amplitudes, and phases of its constituents. Spectrum analysis can be qualitative, for example by visual inspection of the signal, or quantitative, computed from the recorded data. Most often we will use the term *spectrum* to refer to a numerical estimate of the amplitude and phase of the signal at a series of uniformly spaced frequencies. These spectra must be further analyzed, for example by identifying the center frequency or integrating the intensity of a resonance line that spans several frequency values, in order to obtain physically or chemically relevant information. Alternatively, as we shall see in Chapter 4, it is possible to perform parametric spectrum analysis, in which estimates of the spectral parameters are obtained directly from the time series, bypassing altogether the computation of a conventional spectrum.

1.2.2. Resolution and Sensitivity

The terms resolution and sensitivity are used to characterize the quality of a spectrum. *Resolution* refers to the ability to distinguish components of the signal that are close in frequency. The *digital frequency resolution* of a spectrum measures how closely the discrete frequencies making up the spectrum are spaced. The resolution of a spectral estimate and the digital frequency resolution are not the same, although they are clearly related. However, it is often possible to determine the central frequency of a resonance with a precision that exceeds the digital resolution (see Chapter 7).

Sensitivity refers to the ability to distinguish signal from noise. A widely used indicator of sensitivity is the *signal-to-noise ratio* (S/N), which is a number giving the ratio of the highest peak in a spectrum to the height corresponding to the noise level. The noise level is often taken to be the root-mean-square (RMS) value of the noise, that is, the square root of the average of the squared noise values. Sometimes other measures are used; when comparing S/N it is wise to be certain of the definition used for the noise level. While the S/N can be used to quantify the sensitivity, S/N and sensitivity are not the same. For example, we shall see in Chapter 5 that the S/N for spectra obtained by maximum entropy reconstruction is not a good measure of sensitivity. Especially when using nonlinear methods for estimating a spectrum, sensitivity should be assessed explicitly, rather than inferred from the S/N.

1.2.3. Noise

The term *noise* appears repeatedly in this book. In most instances we mean random noise, that is, a statistically random sequence of values with zero mean and (usually) a normal distribution. An important property of random noise is that different members of the sequence are uncorrelated. If ε is the sequence of noise values, these properties can be stated mathematically as

$$\frac{1}{M} \sum_{k=0}^{M-1} \varepsilon_k = 0 \tag{1.1}$$

$$\frac{1}{M} \sum_{k=0}^{M-1} \varepsilon_k^2 = \sigma^2 \tag{1.2}$$

and

$$\sum_{\substack{j,k=0,\\ j \neq k}}^{M-1} \varepsilon_j \varepsilon_k = 0 \tag{1.3}$$

where σ is the RMS value of the noise. Another important property is that when random noise sequences are added element by element, the RMS value increases as the square root of the number of sequences. For two sequences this result is straightforward:

$$\frac{1}{M} \sum_{k=0}^{M-1} (\varepsilon_k + \varepsilon_k')^2 = \frac{1}{M} \sum_{k=0}^{M-1} (\varepsilon_k^2 + 2\varepsilon_k \varepsilon_k' + \varepsilon_k'^2)$$

$$= \sigma^2 + 0 + \sigma^2 = (\sqrt{2}\sigma)^2 \tag{1.4}$$

(Generalization to an arbitrary number of sequences is left as a warm-up exercise for the reader.) There are also many sources of nonrandom noise in NMR experiments. Some of the more systematic varieties are called *artifacts*, and in later chapters we will discuss techniques that have been developed for dealing with them.

1.2.4. The FID and the Spectrum

While the action of various NMR data processing methods can be illustrated using pictures, we must resort to mathematical descriptions in order to convey the details of how the methods work. The starting point for NMR data processing is the FID, which on all modern instruments is sampled at discrete time intervals and recorded as a time series, and can be either real or complex. We denote this series as the vector

$$\mathbf{d} = d_0, d_1, d_2, \dots, d_{M-1} \tag{1.5}$$

where the subscripts indicate the time order of the samples, and M is the total number of samples. Usually, but not always, the interval between samples is constant, so that the time of the kth sample is

$$t_k = k \, \Delta t \tag{1.6}$$

where Δt is the time between samples.

The principal result of NMR data processing is an estimate of the spectrum of $\mathbf{d}$, which we denote as the vector

$$\mathbf{f} = f_0, f_1, f_2, \dots, f_{N-1} \tag{1.7}$$

The subscripts here indicate frequency order, and N is the total number of frequency components in the spectrum estimate. Unlike **d**, **f** is always complex. The interval Δf between frequency components is constant. We prefer to measure frequencies in units of cycles per second (Hz), but they are sometimes measured in radians per second. A frequency f, measured in cycles per second, is converted to the corresponding value ω, in radians per second, by the relation

$$\omega = 2\pi f \tag{1.8}$$

In NMR the signals we wish to measure are caused by nuclear resonances. They show up as *peaks* in the spectrum. Following common practice, we will use the terms ''peak'' and ''line'' interchangeably; thus terms like ''peak shape'' and ''line shape'' both refer to the same thing.

1.3. SOME OLD SAWS: VERY PRACTICAL SPECTRUM ANALYSIS

There are two kinds of old saws—sayings that are repeated so often that they tumble out like some involuntary reaction, and the kind with 10 or 12 teeth to the inch. Like any tool, they can be used indiscriminately, with rather crude results. But when used appropriately, the results can be simple and beautiful. Such are the saws we describe now.

Most of this book describes methods for numerically analyzing the FID, to obtain a spectrum estimate or a set of parameters for a physical model characterizing the sample. However, some very useful spectrum analysis can be performed without a computer, simply by looking at the FID. This qualitative spectrum analysis is invaluable to the spectroscopist for setting up an experiment or tuning a spectrometer, but it is also important to the analyst as a means for deciding what processing steps to apply, or for deciding whether the results of a processing step are reasonable—a kind of zero-order error analysis.

When there is only one predominant frequency component in an FID, the frequency can be estimated by counting the zero-crossings. A signal that decays without crossing zero has a frequency of zero; a signal with lots of zero-crossings has a high frequency, and the more zero-crossings the higher the frequency. The width of a component can be estimated from the rate at which it decays; very narrow resonances decay slowly with time or not at all; broad lines decay very quickly (Fig. 1.3). This is the feature that almost every spectroscopist has exploited at one time or another to ''shim'' the magnetic field on a spectrometer.

When there is more than one strong frequency component present, it is difficult to say much about their characteristics just by looking at the FID, other than that one (or more) of the components must be narrow if the envelope

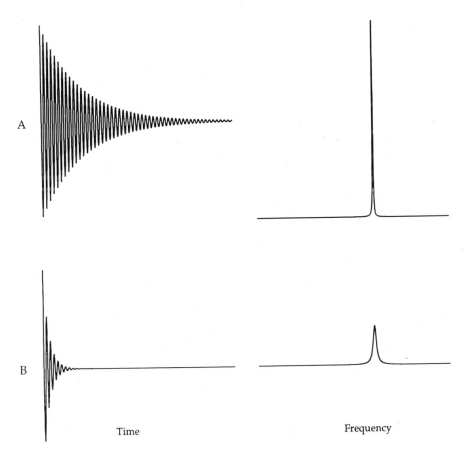

Fig. 1.3. A slowly decaying signal (**A**) gives rise to a narrow line in the spectrum, while a rapidly decaying signal (**B**) gives rise to a broad line in the spectrum.

of the FID decays slowly, and other vague generalities. However, the phenomenon of beating, a consequence of interference, can be used to estimate the *difference* in frequency between two strong components that have similar intensities. Beats appear as a modulation of the intensity of the FID with a frequency given by half the difference in frequency between the interfering components (Fig. 1.4).

Whenever an FID contains more than one strong frequency component, graphical analysis of the FID is nearly useless for characterizing the components. A numerical spectrum estimate that untangles the mutual interference of different frequency components is needed. That's what the rest of this book is about.

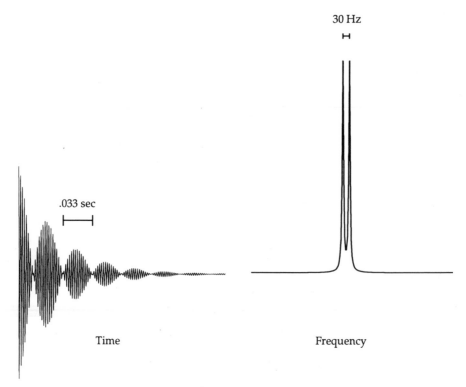

Fig. 1.4. Two components with different frequencies give rise to "beats" in the FID. The duration of each beat is equal to one over the frequency difference.

TO READ FURTHER

Richard Ernst has written a very personal account of the role of computers and NMR data processing in the evolution of modern NMR spectroscopy, complete with vintage photographs [1]. A number of additional historical accounts can be found in the *Encyclopedia of NMR* [2]. A historical account of the development of the FFT is given in the book by Brigham [3]. The year 1965 witnessed the publication of the now-famous paper by Cooley and Tukey [4], and the introduction by Ernst of pulsed Fourier transform NMR at the Sixth Experimental NMR Conference. The published account of the first FT-NMR experiments appeared in 1966 [5].

2

FUNDAMENTALS OF THE DISCRETE FOURIER TRANSFORM

2.1. DEFINITION OF THE DFT

The most important estimate for the spectrum of a time series is given by the discrete Fourier transform,

$$f_n = \frac{1}{\sqrt{N}} \sum_{k=0}^{N-1} d_k e^{-2\pi i k n/N} \qquad (2.1)$$

which yields an N-point spectrum $\mathbf{f}$ from an N-point time series $\mathbf{d}$. The motivation for using the DFT to obtain a spectrum estimate doesn't simply drop out of the blue. It is based on the theory of the continuous Fourier integral (named for Jean-Baptiste Fourier, 1768–1830), linear response theory (which holds that the impulse response of a linear system and the frequency response are related by the Fourier transform), the Wiener-Khinchin theorem, and a whole lot more heavy, formal theory. However, rather than a formal development of the DFT, our treatment will be an investigation of the properties of Eq. (2.1). As we explore these properties we will uncover not only some subtleties, including differences between the continuous and discrete Fourier transforms, but also additional rationales for the use of the DFT as a spectral estimate. Informality reigns; kick off your shoes, loosen your collar, and let's see where it leads us.

It should be noted that the DFT is sometimes defined differently, with other normalization constants and the opposite sign for the argument of the exponential. These differences are mainly just a matter of convention, and have no significant effect on the properties of the DFT as long as they are used consistently.

9

2.2. FREQUENCIES MAKING UP THE DFT SPECTRUM, AND THE SAMPLING THEOREM

If the samples in the time series **d** are uniformly spaced at intervals of Δt seconds, then the frequency components of **f** are spaced at intervals of $1/N\Delta t$ Hz; hence the nth frequency is given by $n/N\Delta t$. Remember the value $1/N\Delta t$; we'll be seeing it a lot. The total width of the spectrum, called the *bandwidth* or the *spectral width* (or simply SW*), is accordingly equal to $1/\Delta t$, since there are N frequencies. The frequencies of the DFT spectrum range from zero to $(N - 1)/N\Delta t$.

Something odd occurs outside this frequency range. Consider what happens if we let n take the value N—that's just one beyond the high-frequency limit. Plugging $n = N$ into Eq. (2.1), we get the same value as for the $n = 0$ component, since $e^{2\pi i Nk/N} = e^{2\pi ik} = 1 = e^{2\pi i 0k/N}$ for every value of k! This is just one manifestation of the more general result that the DFT is periodic in the frequency domain with period $1/\Delta t$. The component at frequency $n/N\Delta t$ is equivalent to the components at frequencies $(n/N\Delta t) \pm (m/\Delta t)$, $m = 1, 2, \ldots$ (Fig. 2.1A).

For many purposes, it is advantageous to consider spectra containing both positive and negative frequency components with zero frequency in the middle, rather than all positive frequencies with zero at one edge. The periodicity of the DFT allows us to take either view. Since frequency components in the range from $1/2\Delta t$ to $1/\Delta t$ are equal to components in the range from $-1/2\Delta t$ to zero, the DFT spectrum can equally well be considered to span the frequencies from $-1/2\Delta t$ to $(N/2 - 1)/N\Delta t$ (Fig. 2.1B and C). We will most often follow this convention for plotting spectra.

A troublesome consequence of the periodic sampling on which the DFT is based is the phenomenon of *aliasing*: Frequencies that differ by multiples of $1/\Delta t$ are completely indistinguishable from one another. The term "aliasing" is apt, because signals with frequencies beyond the limits of the spectrum appear disguised as signals with frequencies inside the limits. This will cause ambiguity, if not outright confusion about the true frequencies present in the signal.

Figure 2.2 illustrates the ambiguity in concrete terms. It shows two sinusoids sampled at intervals of $\Delta t = 0.1$ seconds; the sampled points are indicated by dots. The high-frequency sinusoid has a frequency that is greater than that of the low-frequency sinusoid by $1/\Delta t = 10$ Hz. Since they both take on the same values at the sampled times, they are indistinguishable based on this sampling.

Of course, if it is known beforehand that the signal contains no components with frequencies larger than $1/2\Delta t$ in absolute value, the problem doesn't arise. Such a signal is called *band-limited*, and the well-known sampling theorem

---

*Historically, SW was the *sweep width*, a concept appropriate for frequency-swept continuous wave NMR experiments. Nowadays the two terms are used interchangeably. Fortunately, the abbreviation SW works in either case.

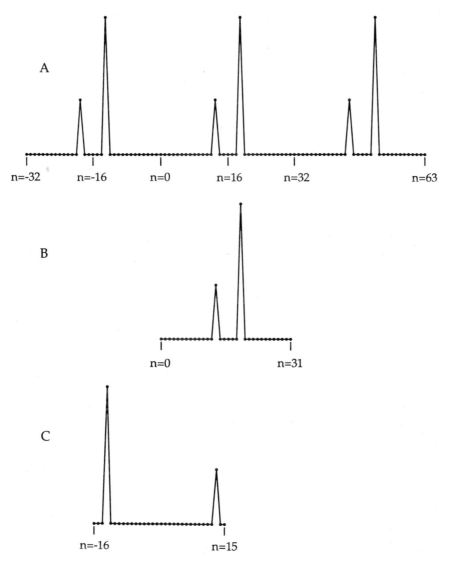

Fig. 2.1. The DFT is implicitly periodic. In this spectrum $N = 32$, but in (**A**) the spectrum has been extended to values of n outside the range 0 to $N - 1$. Points separated by multiples of 32 are all equivalent. Note that the spectrum is equally well represented by the intervals $0 \leq n \leq 31$ and $-16 \leq n \leq 15$, as shown in (**B**) and (**C**).

(also called the Nyquist theorem) states that under these conditions the signal can be completely—and unambiguously—characterized by the time series consisting of its values at discrete intervals Δt. More generally, if all the frequency components of a signal lie within a fixed range of width W, that signal is characterized by any series of samples collected at intervals smaller than the Nyquist interval $1/W$.

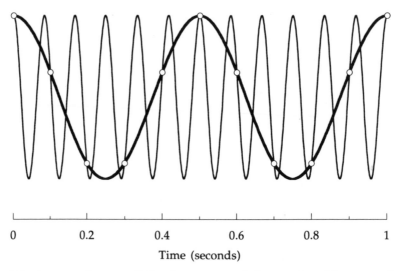

Fig. 2.2. At a sampling rate of 10 points per second, the two sinusoids are completely indistinguishable, even though they have different frequencies (the thick line has a frequency of 2 Hz and the thin line has a frequency of 12 Hz). The open circles represent the sampled data values, which are the same for the two curves. This phenomenon is called *aliasing*.

In practice it rarely happens that the spectrum of a time-varying signal is truly band-limited. It is then important to sample fast enough so that aliasing is negligible, or to condition the signal prior to sampling so as to achieve the same effect. On modern NMR instruments, the spectroscopist chooses the spectral width, and hence the sampling interval Δt. The instrument conditions the signal before sampling by passing it through an analog filter circuit that has a cutoff close to the Nyquist frequency. Noise in particular is almost never band-limited, and this filtering reduces the amount of noise that is aliased into the region of interest, improving the sensitivity.

Aliasing is sometimes referred to as *folding* or *wrapping*, because when frequency components greater than the cutoff frequency are present, they appear at a place in the spectrum that corresponds to folding or wrapping the paper on which the spectrum has been plotted. We will see in the next section that the point about which the spectrum is folded depends not just on the sampling frequency, but also on how the signal is sampled.

2.3. HOW TO SAMPLE: QUADRATURE VERSUS SINGLE-PHASE DETECTION

The presence of the complex exponential in the definition of the DFT [Eq. (2.1)] is a clue that it applies to complex time series and yields a complex spectrum. The implication is that each time the signal is sampled we actually make two measurements, corresponding to the real and imaginary parts of the signal. An oscillating signal can be represented by a vector rotating in a plane, and a detector that measures one component (i.e., X or Y) of the vector is called a *phase-sensitive* detector. When both components are measured, the resulting data can be treated as a complex number, and the process is called *quadrature detection*. Quadrature detection offers a number of advantages over the alternative of merely sampling one component of the signal (called *single-phase detection*).

One advantage is that it enables the DFT to distinguish the sense of rotation, or the sign of the frequency. Figure 2.3 illustrates the X and Y components of signals represented by vectors starting along the X-axis at time zero and rotating at the same frequency but in opposite directions. The X component is the same regardless of the direction of rotation, showing that if all we sampled was the X component we would have no way of distinguishing the sense of rotation. The Y components are different, however. You might ask why we couldn't distinguish the sense of rotation merely by detecting the Y component, rather than the X component. While that would work for the example shown in the figure, it wouldn't work if the two vectors happened to start along the Y-axis.

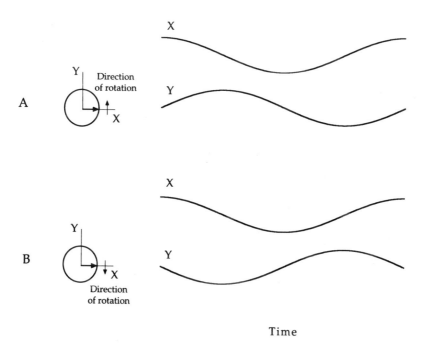

Fig. 2.3. It is possible to determine the sign of a frequency using quadrature detection. Two rotating vectors having the same frequency but opposite directions [counter-clockwise in (**A**) and clockwise in (**B**)] are indistinguishable based on their X components alone. Detection of the Y components allows them to be distinguished.

In general, measurements made along a single axis can't distinguish a motion from its reflection through that axis.

We can see this more explicitly by comparing the DFTs of signals sampled using quadrature and single-phase detection. Consider a signal oscillating with frequency w, aligned with the X-axis at $t = 0$. Suppose we choose a spectral width equal to $4w$; the interval between samples is then $1/4w$. The time series **d** is given by

$$d_k = e^{2\pi i(k/4w)w} = e^{\pi ik/2} = \cos\left(\frac{\pi k}{2}\right) + i \sin\left(\frac{\pi k}{2}\right),$$

$$\mathbf{d} = 1, i, -1, -i, 1, \ldots \tag{2.2}$$

If we compute the DFT of only the real part of **d**, corresponding to single-phase detection, the spectrum is nonzero at *two* frequencies, $\pm w$ (Fig. 2.4A). However, if we compute the DFT of the full complex time series, corresponding to quadrature detection, the result is nonzero only at w (Fig. 2.4B).

Quadrature detection has an additional advantage. Usually the noise in the two components is statistically independent, so the S/N is larger by a factor of

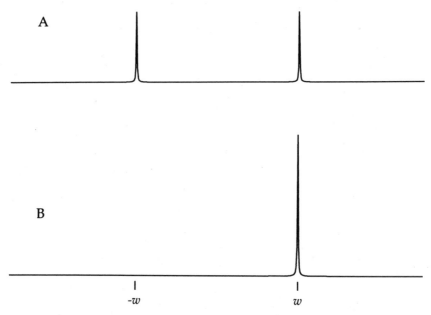

Fig. 2.4. The DFT of the real part (single-phase detected) of a single sinusoid (**A**) includes two frequency components that are opposite in sign, while the DFT of the full complex (quadrature detected) sinusoid (**B**) contains only one component at the true frequency of the sinusoid.

$\sqrt{2}$ than it is for single-phase detection. This is simply because twice as many data values are available.

2.4. DIGITAL RESOLUTION AND THE BANDWIDTH–RESOLUTION TRADE-OFF

So far we have considered the influence of the rate of sampling on the DFT. Another characteristic that distinguishes the DFT from its continuous cousin is that it is not only discrete, but finite as well: The number of samples influences the properties of the DFT. The spacing between frequencies, or digital resolution, of the DFT is $1/N\Delta t$. If the number of sample points N is fixed, then increasing the spectral width of the DFT (by sampling more rapidly—decreasing Δt) worsens the digital resolution; it increases the spacing between frequencies. Of course the reverse is also true: Decrease the spectral width and the digital resolution improves.

2.5. THE INVERSE DFT

Given the complex N-point spectrum $\mathbf{f}$, it is possible to compute the time series $\mathbf{d}$ that gives rise to the spectrum (provided it is appropriately band-limited) by using the inverse discrete Fourier transform (IDFT):

$$d_k = \frac{1}{\sqrt{N}} \sum_{n=0}^{N-1} f_n e^{2\pi ikn/N} \tag{2.3}$$

In essence, this corresponds to adding together a collection of single-frequency signals $e^{2\pi ikn/N}$, each scaled by the appropriate amplitude and phase f_n.

It's easy to show that Eq. (2.3) describes a true inverse of the DFT, that is, $\mathbf{d} = \text{IDFT}[\text{DFT}(\mathbf{d})]$. Substituting the right-hand side of Eq. (2.1) into Eq. (2.3) we get

$$\text{IDFT}[\text{DFT}(\mathbf{d})]_k = \frac{1}{\sqrt{N}} \sum_{n=0}^{N-1} \left(\frac{1}{\sqrt{N}} \sum_{k'=0}^{N-1} d_{k'} e^{-2\pi ik'n/N} \right) e^{2\pi ikn/N}$$

$$= \frac{1}{N} \sum_{k'=0}^{N-1} \left(\sum_{n=0}^{N-1} e^{2\pi i(k-k')n/N} \right) \tag{2.4}$$

The sum over n is equal to N for $k = k'$ and zero otherwise. This is the *orthogonality* property of complex exponentials, which we will demonstrate shortly. The right-hand side of Eq. (2.4) thus reduces to d_k, showing that Eq.

(2.3) is indeed the inverse of Eq. (2.1). The converse proof that DFT[IDFT($\mathbf{f}$)] = $\mathbf{f}$ is essentially the same.

The orthogonality of complex exponentials is proved as follows. Since

$$e^{2\pi i(k-k')0/N} = e^{2\pi i(k-k')N/N} = 1 \tag{2.5}$$

we have

$$\sum_{n=0}^{N-1} e^{2\pi i(k-k')n/N} = \sum_{n=1}^{N} e^{2\pi i(k-k')n/N}$$

$$= \sum_{n=0}^{N-1} e^{2\pi i(k-k')(n+1)/N} \tag{2.6}$$

$$= e^{2\pi i(k-k')/N} \sum_{n=0}^{N-1} e^{2\pi i(k-k')n/N}$$

and therefore

$$[1 - e^{2\pi i(k-k')/N}] \sum_{n=0}^{N-1} e^{2\pi i(k-k')n/N} = 0 \tag{2.7}$$

Since $k \neq k'$ and both lie in the range from zero to $N-1$, $(k-k')$ is not divisible by N and so $e^{2\pi i(k-k')/N}$ is not equal to one. For the product in Eq. (2.7) to be equal to zero, it must be true that

$$\sum_{n=0}^{N-1} e^{2\pi i(k-k')n/N} = 0 \tag{2.8}$$

2.6. MATRIX REPRESENTATION OF THE DFT

The DFT can be represented as an ($N \times N$) matrix $\mathbf{F}$, which operates on the data vector to yield the spectrum:

$$\mathbf{f} = \mathbf{Fd} \tag{2.9}$$

Similarly, the IDFT can be represented as a matrix $\mathbf{F}'$. Comparison of the definitions of the DFT [Eq. (2.1)] and the IDFT [Eq. (2.3)] shows that the elements of $\mathbf{F}'$ are related to those of $\mathbf{F}$ by

$$F'_{nk} = F^*_{kn} \tag{2.10}$$

That is, the matrix $\mathbf{F}'$ is the complex conjugate of the transpose of $\mathbf{F}$. The complex conjugate transpose of a matrix $\mathbf{M}$ is called the *Hermitian transpose* (or the *Hermitian conjugate*), denoted $\mathbf{M}^\dagger$. Since the IDFT is the inverse of the DFT, we have

$$\mathbf{d} = \mathbf{F}^\dagger \mathbf{F} \mathbf{d} \qquad (2.11)$$

which means that $\mathbf{F}^\dagger \mathbf{F} = \mathbf{I}$, the identity matrix. Therefore $\mathbf{F}^\dagger$ is just the matrix inverse of $\mathbf{F}$. A matrix whose inverse is equal to its Hermitian transpose is called *unitary*.

The columns of $\mathbf{F}$ can be interpreted as individual vectors. Using the results of the previous section, it is easy to show that these vectors are mutually orthogonal and are normalized, that is, $\sum_{n=0}^{N-1} v_n^* w_n$ is zero if $\mathbf{v}$ and $\mathbf{w}$ are distinct column vectors, and one if they are the same column vector. This means that the column vectors form an orthonormal basis for the space of N-dimensional complex vectors; it is sometimes referred to as the *Fourier basis*. The DFT can thus be viewed as a change of basis; for any data vector $\mathbf{d}$, DFT($\mathbf{d}$) is the vector giving the coefficients of $\mathbf{d}$ expanded in this new basis.

2.7. SPECIAL POINTS

It is possible to attach special significance to certain points of the vectors $\mathbf{f}$ and $\mathbf{d}$. The defining equation (2.3) shows that the first point of the time series, d_0, is equal to the sum of all the frequency components divided by $\sqrt{N}$. Similarly, Eq. (2.1) shows that the first point of the spectrum, f_0, which is the zero-frequency or "DC" component, is equal to the sum of the time series divided by $\sqrt{N}$.

2.8. PARSEVAL'S THEOREM

For any signal, the power concentrated at a particular frequency is given by the squared magnitude of that frequency component in the spectrum. Accordingly, the total power contained in a spectrum is the sum of the squared magnitudes of all the frequency components:

$$\text{total power} = \sum_{n=0}^{N-1} |f_n|^2 \qquad (2.12)$$

Using either Eq. (2.1) or Eq. (2.3), it's easy to show (even for us!) that

$$\sum_{k=0}^{N-1} |d_k|^2 = \sum_{n=0}^{N-1} |f_n|^2 \qquad (2.13)$$

Here is the proof:

$$\sum_{n=0}^{N-1} |f_n|^2 = \sum_{n=0}^{N-1} f_n f_n^*$$

$$= \sum_{n=0}^{N-1} \left(\frac{1}{\sqrt{N}} \sum_{k=0}^{N-1} d_k e^{-2\pi ikn/N} \right) \left(\frac{1}{\sqrt{N}} \sum_{k'=0}^{N-1} d_k^* e^{2\pi ik'n/N} \right)$$

$$= \sum_{k=0}^{N-1} \sum_{k'=0}^{N-1} d_k d_{k'}^* \left(\frac{1}{N} \sum_{n=0}^{N-1} e^{-2\pi i(k-k')n/N} \right) \tag{2.14}$$

$$= \sum_{k=0}^{N-1} d_k d_k^* = \sum_{k=0}^{N-1} |d_k|^2$$

where the orthogonality of the complex exponentials is used to eliminate the sum over n. This important result is known as Parseval's theorem; it says that the power is the same whether computed in the time or frequency domain. [Here is a case where the conventions make a difference. One common convention is to use 1 as the normalization constant for the DFT and $1/N$ for the IDFT. Parseval's theorem would then state that power($\mathbf{f}$) = $N \cdot$ power($\mathbf{d}$).]

2.9. FREQUENCY SHIFTING: DEMODULATION

Consider what happens to the time series $\mathbf{d}$ if we shift its DFT spectrum to the left by an integral number of points j. The new spectrum $\mathbf{f}'$ is given by $f_n' = f_{n+j}$, and since the spectrum $\mathbf{f}$ is periodic the shift is circular: $f_{N-j}' = f_0$, and so on. We substitute into Eq. (2.1):

$$f_n' = f_{n+j} = \frac{1}{\sqrt{N}} \sum_{k=0}^{N-1} d_k e^{-2\pi ik(n+j)/N}$$

$$= \frac{1}{\sqrt{N}} \sum_{k=0}^{N-1} (d_k e^{-2\pi ikj/N}) e^{-2\pi ikn/N} \tag{2.15}$$

Hence shifting the spectrum by j points is equivalent to multiplying the time domain signal by $e^{-2\pi ikj/N}$. This process is called *demodulation*. In addition to providing a useful tool, this result also suggests another way of interpreting the DFT: Each component f_n of the DFT spectrum represents the zero-frequency or "DC" component after demodulating the signal by the frequency $n/N\Delta t$.

Suppose we want to shift the spectrum by an amount s that isn't an integral multiple of $1/N\Delta t$. We just multiply each point d_k in the time domain signal by $e^{-2\pi iks\Delta t}$. For an arbitrary shift, however, the relationship between the DFT of the result and the DFT of the original time series is not obvious or straightforward, since we are now effectively interpolating between points of the original spectrum. It might appear at first glance that demodulation is a way to obtain a spectrum with arbitrarily high digital resolution: Any point in the spectrum can be found by demodulating the time series by the corresponding amount. The accuracy of this interpolation, however, depends very much on the nature of the signal, as we will see in the section on leakage.

2.10. TIME SHIFTING: LINEAR PHASE SHIFTING

Suppose we shift the signal in the time domain instead of the frequency domain. Using Eq. (2.3), the result looks eerily familiar:

$$d_{k-j} = \frac{1}{\sqrt{N}} \sum_{n=0}^{N-1} f_n e^{2\pi i n(k-j)/N}$$
$$= \frac{1}{\sqrt{N}} \sum_{n=0}^{N-1} (f_n e^{-2\pi i n j/N}) e^{2\pi i k n/N} \qquad (2.16)$$

This says that shifting the time domain signal by j points (also in the circular sense) is the same as multiplying the DFT of the original signal by $e^{-2\pi i n j/N}$; this process is called *linear phase shifting*. Multiplying a complex number by an imaginary exponential $e^{-2\pi i n j/N}$ results in a phase shift (i.e., a rotation in the complex plane) of $-2\pi n j/N$ radians; this is a linear phase shift because the amount of rotation increases—linearly—with the frequency. A shift of one point in the time domain corresponds to a linear phase shift by an amount that varies from zero at zero frequency to $[(N - 1)/N]2\pi$ radians at the point $N - 1$ in the spectrum. This is commonly referred to simply as a 360° linear phase shift, because over the width of the spectrum the phase shift changes by 360°.

2.11. LINEARITY

Another result that follows immediately from Eq. (2.1) is the linearity of the DFT. Linearity means that the DFT of the sum of two time series is equal to the sum of the DFTs of the individual time series. More generally,

$$\text{DFT}(a\mathbf{d} + b\mathbf{d}') = a\mathbf{f} + b\mathbf{f}' \qquad (2.17)$$

where $\mathbf{d}$ and $\mathbf{d}'$ are the time series, a and b are constants, and $\mathbf{f}$ and $\mathbf{f}'$ are the DFTs of $\mathbf{d}$ and $\mathbf{d}'$, respectively. But hold on. If you are slouching down in your chair, you might want to sit up and pay attention to this. Just because this result is obvious doesn't mean it's not important. Linearity means that twice as much signal yields twice as much frequency response. There are few properties that are more important for the use of the DFT in quantitative analysis.

2.12. SYMMETRY PROPERTIES

Symmetry can variously be fear-inspiring (Blake's tiger), beautiful (a flower), or merely useful (a hex nut). Some people would claim that the DFT possesses all three qualities. Be that as it may, our main concern here is its utility. DeMoivre's formula,

$$e^{iz} = \cos z + i \sin z \qquad (2.18)$$

tells us that the complex exponentials that permeate the DFT are really trigonometric functions in disguise. It should come as no surprise, therefore, that the DFT is loaded with symmetries. A concrete example of their utility is the class of FFT algorithms, which exploit these symmetries in the extreme. While development of FFT algorithms is beyond the scope of this book, some symmetries of the DFT are nevertheless quite useful in quotidian NMR data processing.

One of the more useful symmetries of the DFT is reversal symmetry. But first we have to be careful about what we mean by reversal. A time series is reversed by changing the sign of each index, which by the implicit periodicity of the DFT means that d_k is exchanged with d_{N-k}, for $k > 0$. Note that d_0 remains unaffected by this operation, so that reversal is not quite the same as reversing the order of elements in the vector $\mathbf{d}$. A similar consideration applies to reversal in the frequency domain.

Let $\mathbf{f}'$ be the DFT of the reversal of the time series $\mathbf{d}$. We take our usual approach by replacing d_k with d_{N-k} in Eq. (2.1),

$$f'_n = \frac{1}{\sqrt{N}} \sum_{k=0}^{N-1} d_{N-k} e^{-2\pi i k n / N} \qquad (2.19)$$

Letting $k' = N - k$ gives

$$f'_n = \frac{1}{\sqrt{N}} \sum_{k'=N}^{1} d_{k'} e^{-2\pi i (N - k') n / N} \qquad (2.20)$$

Since $e^{-2\pi i N n / N} = e^{-2\pi i 0 n / N}$ and $d_0 = d_N$ (implicit periodicity), we have

$$f'_n = \frac{1}{\sqrt{N}} \sum_{k'=0}^{N-1} d_{k'} e^{-2\pi i k' (-n) / N} \qquad (2.21)$$

which is the same as

$$f'_n = \frac{1}{\sqrt{N}} \sum_{k'=0}^{N-1} d_{k'} e^{-2\pi i k' (N - n) / N} \qquad (2.22)$$

showing that $f'_n = f_{N-n}$. Thus reversing the time domain data is equivalent to reversing the DFT spectrum.

Another useful symmetry is a change of sign of the imaginary part: complex conjugation. You know the drill.

$$
\begin{aligned}
f'_n &= \frac{1}{\sqrt{N}} \sum_{k=0}^{N-1} d_k^* e^{-2\pi i k n/N} \\
&= \frac{1}{\sqrt{N}} \sum_{k=0}^{N-1} (d_k e^{2\pi i k n/N})* \\
&= \frac{1}{\sqrt{N}} \sum_{k=0}^{N-1} (d_k e^{-2\pi i k(-n)/N})* \\
&= \frac{1}{\sqrt{N}} \sum_{k=0}^{N-1} (d_k e^{-2\pi i k(N-n)/N})* \\
&= (f_{N-n})*
\end{aligned}
\tag{2.23}
$$

So changing the sign of the imaginary part of the time domain data reverses the spectrum and changes the sign of its imaginary part.

To give an example of when these symmetries might come in handy, suppose you are stranded on a desert isle with a computer program that computes only the forward DFT, but you need to compute an inverse DFT. No problem:

$$
\begin{aligned}
d_k &= \frac{1}{\sqrt{N}} \sum_{n=0}^{N-1} f_n e^{2\pi i k n/N} \\
&= \frac{1}{\sqrt{N}} \sum_{n'=N}^{1} f_{N-n'} e^{2\pi i k(N-n')/N} \\
&= \frac{1}{\sqrt{N}} \sum_{n'=0}^{N-1} f_{N-n'} e^{-2\pi i k n'/N}
\end{aligned}
\tag{2.24}
$$

This shows that the inverse DFT of **f** is the same as the DFT of the reverse of **f** (or, if you prefer, the reverse of the DFT of **f**). Voilà!

Other symmetries follow from these. Suppose the time domain data are purely real or purely imaginary. For purely real data, $d_k = d_k^*$. Therefore $f_n = (f_{N-n})^*$, which means that the real part of **f** has even symmetry:

$$
\text{real}(f_n) = \text{real}(f_{N-n})
\tag{2.25}
$$

and the imaginary part has odd symmetry:

$$
\text{imag}(f_n) = -\text{imag}(f_{N-n})
\tag{2.26}
$$

It is easy to see that if the data **d** are purely imaginary then the real part of **f** has odd symmetry and the imaginary part has even symmetry.

We have already mentioned that a different sign convention is often used for the argument of the exponential of the DFT. The DFT according to that convention is what we are calling the IDFT, and vice versa. From the discussion above, we can see that the two conventions yield spectra that are reversed with respect to each other.

2.13. CONVOLUTION

Another result worth sitting up and paying attention to, though one that is not quite so obvious as linearity, is the effect of multiplying the time domain signal, point by point, by a periodic but otherwise arbitrary vector **a**, with elements a_k. The convolution theorem states that the DFT of the product of the vectors **d** and **a** is given by the *convolution* of the DFT of **d** with the DFT of **a**:

$$\text{DFT}(\mathbf{da})_n = \frac{1}{\sqrt{N}} \sum_{j=0}^{N-1} f_j A_{n-j} \tag{2.27}$$

where **A** is the DFT of **a** and the subscript $n - j$ wraps around in the usual circular manner. The proof is straightforward. Substituting the definition of the DFT into Eq. (2.27) gives

$$\sum_{j=0}^{N-1} f_j A_{n-j} = \frac{1}{N} \sum_{j=0}^{N-1} \left(\sum_{k=0}^{N-1} d_k e^{-2\pi i k j/N} \right) \left(\sum_{k'=0}^{N-1} a_{k'} e^{-2\pi i k'(n-j)/N} \right)$$

$$= \frac{1}{N} \sum_{k=0}^{N-1} \sum_{k'=0}^{N-1} \left[d_k a_{k'} e^{-2\pi i k'n/N} \left(\sum_{j=0}^{N-1} e^{-2\pi i k j/N} e^{2\pi i k'j/N} \right) \right] \tag{2.28}$$

The last part in parentheses reduces to N when $k = k'$, and zero otherwise, due to orthogonality. We are left with

$$\sum_{j=0}^{N-1} f_j A_{n-j} = \sum_{k=0}^{N-1} d_k a_k e^{-2\pi i k n/N} \tag{2.29}$$

which is $\sqrt{N}$ times the DFT of the product of **a** and **d**.

Equation (2.27) is the so-called frequency convolution theorem. There is an analogous time convolution theorem which states that

$$\text{IDFT}(\mathbf{fA})_k = \frac{1}{\sqrt{N}} \sum_{j=0}^{N-1} d_j a_{k-j} \tag{2.30}$$

We're getting a bit bored, so prove this one yourself. (Take heart; the proof is nearly the same.)

The convolution theorem describes one of the more important properties of the DFT, and it serves as the basis for enhancing spectra by weighting the data in the time domain. We'll discuss some of the applications of the convolution theorem in the next chapter.

2.14. THE POWER SPECTRUM

The *power spectrum* $\mathbf{P}$ is defined as the squared magnitude of the spectrum: $P_n = |f_n|^2$. Since $|f_n|^2 = f_n f_n^*$, and taking the complex conjugate of $\mathbf{f}$ is equivalent to reversing and conjugating $\mathbf{d}$, the convolution theorem tells us that $\mathbf{P}$ is the DFT of $\mathbf{d}$ convolved with its reversed complex conjugate:

$$\mathbf{P} = \mathrm{DFT}(\mathbf{p}) \quad \text{where} \quad p_k = \frac{1}{\sqrt{N}} \sum_{j=0}^{N-1} d_j d_{j-k}^* \qquad (2.31)$$

The time series $\mathbf{p}$ is called the *autocorrelation* of $\mathbf{d}$. This result is the famous Wiener-Khinchin theorem.

2.15. CAUSALITY AND THE HILBERT TRANSFORM

Using the convolution theorem, we can demonstrate a more subtle symmetry of the DFT that emerges when the time domain data satisfy a condition characteristic of the response of linear systems, known as *causality*. Causality simply says that a response can't precede its stimulus. This means that if the impulse is applied as a delta-function at $t = 0$, the response at times $t < 0$ must be zero. (Causal signals should not be confused with casual signals, which are another beast altogether.)

This creates a dilemma, since the periodicity of the DFT prevents the data from being zero for $t < 0$! The only way to preserve causality in the discrete case, then, is to sample for equal times before and after the stimulus. Or, if we really believe the response is causal, we can simply precede the measured data with an equal number of zeros, or equivalently, add an equal number of zeros after the sampled data. We will show that when a time series has arbitrary values d_k for $k = 0, 1, \ldots, N/2 - 1$ and is zero for $k = N/2, N/2 + 1, \ldots, N - 1$, there is a special relationship between the real and imaginary components of the DFT spectrum.

Assume that $\mathbf{d}$ is causal, and let

$$\mathbf{a} = \mathrm{IDFT}[\mathrm{real}(\mathbf{f})] \quad \text{and} \quad \mathbf{b} = \mathrm{IDFT}[\mathrm{imag}(\mathbf{f})] \qquad (2.32)$$

Since $\mathbf{f} = \text{real}(\mathbf{f}) + i \, \text{imag}(\mathbf{f})$, by the linearity of the IDFT it follows that $\mathbf{d} = \mathbf{a} + i\mathbf{b}$. Since $\mathbf{a}$ and $\mathbf{b}$ are inverse Fourier transforms of real vectors,

$$a_{N-k} = a_k^* \quad \text{and} \quad b_{N-k} = b_k^* \tag{2.33}$$

For $0 < k \le N/2$, we have

$$0 = d_{N-k} = a_{N-k} + ib_{N-k} = a_k^* + ib_k^* \tag{2.34}$$

and also

$$d_k = a_k + ib_k \tag{2.35}$$

The solution to these simultaneous equations is

$$a_k = ib_k = d_k/2$$
$$a_{N-k} = -ib_{N-k} = d_k^*/2 \tag{2.36}$$

We can write this as

$$b_k = h_k a_k \tag{2.37}$$

where

$$h_k = \begin{cases} -i & 0 < k \le \dfrac{N}{2} - 1 \\[2mm] i & \dfrac{N}{2} \le k \le N - 1 \end{cases} \tag{2.38}$$

The convolution theorem tells us the relationship between the real and imaginary parts of $\mathbf{f}$ is

$$\text{imag}(f_n) = \frac{1}{\sqrt{N}} \sum_{j=0}^{N-1} \text{real}(f_j) H_{n-j} \tag{2.39}$$

where $\mathbf{H}$ is the DFT of $\mathbf{h}$. Equation (2.39) is called the *discrete Hilbert transform*. It shows that the imaginary part of the spectrum of a causal signal can be obtained from the real part by convolution. If you are particularly keen-eyed, however, you may have noticed that the status of h_0 is somewhat cloudy. Since the reversal symmetry operation maps a_0 to itself (and b_0 to itself), Eq. (2.33) tells us that a_0 and b_0 are both real, but the relation between them is otherwise undefined. In particular, it is impossible to determine the value of

b_0 from the value of a_0. The first point of a time series determines the "DC" level (or offset) of the spectrum; consequently the imaginary part of the spectrum of a causal sequence can only be determined from the real part up to an arbitrary offset.

For very large multidimensional spectra it is often useful to conserve computer memory by discarding the imaginary part; if phase mixing needs to be performed, the imaginary component can be reconstructed via the discrete Hilbert transform. In practice it is usually less work to use the convolution theorem to compute the Hilbert transform (apply the inverse DFT, multiply by **h**, and DFT) than to compute the convolution directly. The only caveat is that care should be taken to align the baseline of the imaginary part appropriately.

A somewhat more practical consequence of causality is that the real and imaginary parts of the spectrum do not carry independent information. We are therefore fully justified in presenting only the real part, as is common practice. In fact, data should normally be zero-filled by at least a factor of two, to ensure that the real part of the spectrum by itself is capable of realizing the full benefit of the information present.

2.16. LEAKAGE

We have repeatedly emphasized that the definition of the DFT treats both the discrete, finite data and the similarly discrete and band-limited spectrum as being periodic. But what if the underlying signal is *not* periodic, or has components whose periods do not evenly divide $N\Delta t$? Then the signal contains frequencies that do not correspond exactly with the discrete frequency values of the DFT spectrum, that is, with integer multiples of $1/N\Delta t$. The DFT has difficulty characterizing such frequencies accurately, and some intensity "leaks" to nearby frequencies to compensate. The result is called *leakage*, and it is illustrated in Figure 2.5. Figure 2.5A shows the DFT of an exponentially decaying sinusoid that has a frequency corresponding to an integer multiple of $1/N\Delta t$. Figure 2.5B shows the DFT of an otherwise identical signal whose frequency differs by one half of $1/N\Delta t$; the resulting distortion is sometimes confused with a phase error, since it has the same sort of appearance. Figure 2.5C shows that this effect only manifests itself when the line width is comparable to $1/N\Delta t$; for broader lines the distortion is hardly noticeable.

The most common source of leakage is a discontinuity in the data. A discontinuity gives rise to a continuous (and not band-limited) set of frequencies. The discrete and finite set of frequencies making up the DFT spectrum is not well suited to characterize such signals, and the result, again, is leakage. Figure 2.6 illustrates the influence of one common type of discontinuity: truncation. The leakage caused by truncation can create considerable confusion in spectra that contain many components, and methods for minimizing this leakage con-

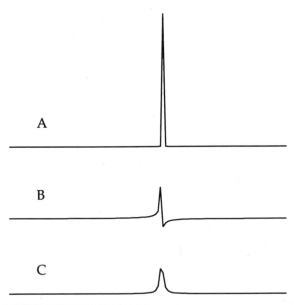

Fig. 2.5. The DFT of a signal with frequency equal to an integer multiple of $1/N\Delta t$ (**A**) shows no leakage artifacts. The DFT of a sinusoid that differs in frequency by $1/2N\Delta t$ from the signal in (**A**) exhibits noticeable leakage (**B**). The DFT of the same signal as (**B**), but with a faster decay rate, shows that the leakage does not manifest when the line width is much larger than $1/N\Delta t$ (**C**).

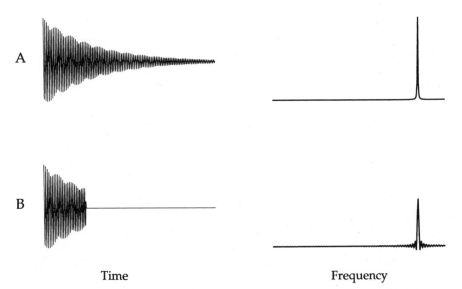

Time   Frequency

Fig. 2.6. Truncation gives rise to artifacts in the DFT spectrum. The original untruncated signal (**A**) has a DFT spectrum with no artifacts, while the truncated signal (**B**) yields a DFT spectrum showing typical leakage artifacts. The extra peaks in (**B**) surrounding the main peak are called "sinc wiggles."

stitute an important part of NMR data processing. We will discuss them at length in Chapter 3.

2.17. DFT OF SPECIAL FUNCTIONS

A number of special functions appear quite often in NMR, and a knowledge of the characteristics of their Fourier transforms will prove useful.

2.17.1. Delta Function

A discrete delta function is one that is zero everywhere except for a single point (Fig. 2.7A). We can see from Eq. (2.1) that if the first point of the time series is the only nonzero point, the DFT is just a constant, $1/\sqrt{N}$. When the nonzero point is not the first point, the result is a single sinusoid snaking its way across the spectrum; the number of periods of the sinusoid is equal to k, the index of the nonzero data point (Fig. 2.7B). If k is small, the slow oscillation in the spectrum is similar to the curved baseline that sometimes occurs in NMR spectra. Not surprisingly, baseline curvature can generally be traced to imperfections in the early part of the FID.

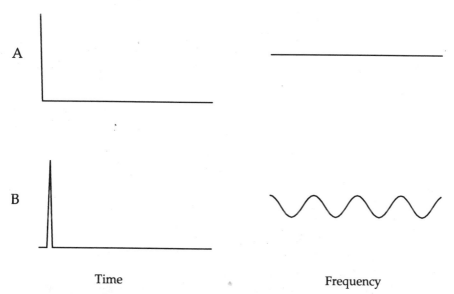

Time   Frequency

Fig. 2.7. The DFT spectrum of a "delta" function whose value is equal to one at the first point and equal to zero everywhere else (**A**) is contant. The DFT spectrum of a delta function that is nonzero at point k in the time domain (**B**) consists of a sinusoid that has k periods. Here $k = 4$.

2.17.2. Step Function

A step function is one that is constant for part of the way, and zero for part of the way, with a single transition between the two values (Fig. 2.8A). Step functions are useful because functions with a single jump discontinuity can be represented as the sum of a continuous function and a step function. The DFT of a step function depends very much on where the transition occurs. It's not so easy to see this by substituting a step function into Eq. (2.1), so we will illustrate the point with a few examples. Figure 2.8 shows the real parts of the DFT spectra of several step functions that differ in the position of the "step." The DFT of a step function also depends fairly dramatically on the total number of points, as can be seen in Figure 2.8B and C.

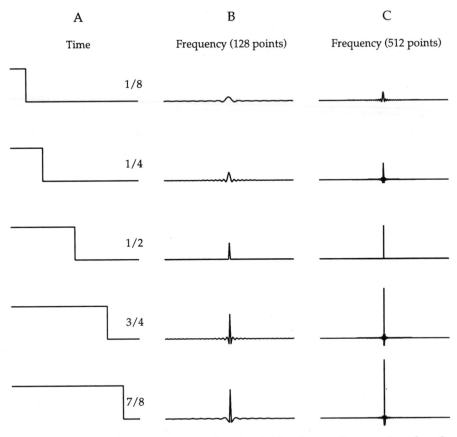

Fig. 2.8. The DFT spectrum of a step function (with value equal to one from $k = 0$ to J, and equal to zero everywhere else) consists of a peak centered at zero frequency together with sinc wiggles. The appearance of the sinc wiggles depends on the value of J and the number of points in the sample N. (**B**) $N = 128$. (**C**) $N = 512$. For $J = N/2$ (the middle row), sinc wiggles do not appear in the real part of the spectrum, but they do appear in the imaginary part (not shown).

The continuous Fourier transform offers additional insight into the shape of the spectrum of a step function. The continuous transform of a step function that is one from zero to S is:

$$f(w) = \int_{-\infty}^{+\infty} d(t)e^{-2\pi iwt}\, dt = \int_{0}^{S} e^{-2\pi iwt}\, dt$$

$$= \frac{i}{2\pi w}\, (e^{-2\pi iwS} - 1) \tag{2.40}$$

$$= \frac{\sin 2\pi wS}{2\pi w} + i\, \frac{\cos 2\pi wS - 1}{2\pi w}$$

The real part is a scaled version of the ubiquitous sinc function: $\sin(\pi x)/\pi x$. The width of the step, S, affects the width of the oscillations. Discontinuities in the data always manifest as "sinc wiggles" in the DFT spectrum.

2.17.3. Exponential Decay

In theory, the FID in pulsed NMR experiments consists of exponentially decaying sinusoids. As a simple example, we will consider a sinusoid that has zero frequency, that is, just an exponential decay with rate πL. The continuous Fourier transform again provides the quickest route to the form of the corresponding spectrum, the so-called *Lorentzian* line shape (see Fig. 2.9):

$$f(w) = \int_{0}^{\infty} e^{-\pi Lt}\, e^{-2\pi iwt}\, dt = \int_{0}^{\infty} e^{-(\pi L + 2\pi iw)t}\, dt$$

$$= \frac{1}{(\pi L + 2\pi iw)} \tag{2.41}$$

$$= \frac{\pi L}{(\pi L)^2 + (2\pi w)^2} - i\, \frac{2\pi w}{(\pi L)^2 + (2\pi w)^2}$$

The maximal value of the real part is $1/\pi L$, which occurs at $w = 0$. Half this maximal value occurs at $w = \pm L/2$. For this reason L is called the full width of the peak at half maximum, or FWHM. (Occasionally one sees the term HWHM, which stands for half width at half maximum.)

Up to now we've avoided looking at the imaginary parts of spectra, but here is a perfect opportunity to do so. Since Lorentzians occur so frequently in NMR, and the real and imaginary parts are often mixed, it's important to be familiar with the characteristics of both. Figure 2.9 shows the real and imaginary parts of a Lorentzian superimposed. The main thing to notice is the broad "tails" (or "wings") of the imaginary part. In a crowded spectrum, these broad tails will overlap with other peaks, whereas the real parts will not overlap

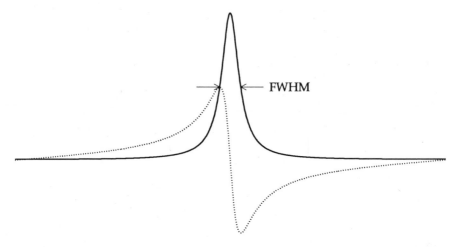

Fig. 2.9. The continuous Fourier transform of a decaying exponential is the Lorentzian line shape. The solid line shows the real part and the dotted line shows the imaginary part. The arrows indicate the full width at half maximum.

nearly as much. This is another reason for displaying only the real parts of complex spectra. The relatively narrow, positive, and symmetric line shape of the real part is called *absorptive*; the broad, antisymmetric shape of the imaginary part is called *dispersive*.

In the discrete case, exponential decay leads to the same line shape—almost. Substituting an exponential decay (with zero frequency) into Eq. (2.1) gives

$$f_n = \frac{1}{\sqrt{N}} \sum_{k=0}^{N-1} e^{-\pi L k \Delta t} e^{-2\pi i k n/N} = \frac{1}{\sqrt{N}} \sum_{k=0}^{N-1} z^k \qquad (2.42)$$

where $z = e^{-\pi L \Delta t - 2\pi i n/N}$. This is just a geometric series, which in closed form is equal to

$$f_n = \frac{1}{\sqrt{N}} \frac{1 - z^N}{1 - z} = \frac{1}{\sqrt{N}} \frac{1 - e^{-\pi L N \Delta t - 2\pi i n}}{1 - e^{-\pi L \Delta t - 2\pi i n/N}} \qquad (2.43)$$

When $\pi L N \Delta t$ is large, that is, when the decay is rapid compared to the length of the sample (or, equivalently, the line width is large compared to the digital resolution), the z^N term approaches zero and we obtain

$$f_n = \frac{1}{\sqrt{N}} \frac{1}{1 - e^{-\pi L \Delta t - 2\pi i n/N}} \qquad (2.44)$$

This clearly is not the same as the continuous case, but as can be seen in Figure 2.10, the two forms are very close. The major difference is a constant vertical offset relative to the continuous Lorentzian (visible in Fig. 2.10A). The size of the offset can be determined by evaluating f_n for $n = N/2$ (that is, the point in the spectrum farthest from the peak, corresponding to the edge of the center column of Fig. 2.10). If $\pi L \Delta t$ is close to zero (the line width is small compared to the spectral width), Eq. (2.44) reduces to $1/2\sqrt{N}$, whereas Eq. (2.41) approaches zero. This offset is not visible in Figure 2.10B and C because the scales of the plots are different.

When $\pi L N \Delta t$ is small (the signal has not decayed to zero by the end of the sampling period), the z^N term cannot be ignored. It accounts for the large discrepancy in peak heights shown in Figure 2.10C. Intuitively, it is easy to understand why the discrete Lorentzian should be smaller, since much of the signal lies beyond the end of the sampling period.

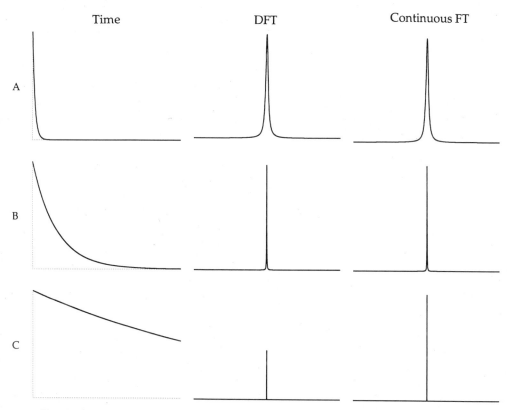

Fig. 2.10. The discrete and continuous Fourier transforms of a decaying exponential are almost the same. (**A**), (**B**), and (**C**) show signals with differing decay rates; the spectra in the three rows are not to the same scale (although the two spectra on each row are to scale). The small baseline offset visible in (**A**) has consequences for multidimensional spectra (discussed in Chapter 3).

What happens if the exponentially decaying sinusoid has a frequency w_0 other than zero? In the time domain, the exponential decay multiplies the sinusoid:

$$d(t) = e^{-\pi Lt}e^{2\pi i w_0 t} \tag{2.45}$$

The continuous Fourier transform yields a Lorentzian line centered at w_0:

$$f(w) = \frac{\pi L}{(\pi L)^2 + (2\pi(w - w_0))^2} - i \frac{2\pi(w - w_0)}{(\pi L)^2 + (2\pi(w - w_0))^2} \tag{2.46}$$

The convolution theorem allows us to understand the spectrum of a decaying sinusoid as the convolution of a delta function at w_0 (the Fourier transform of the sinusoid) with a Lorentzian of width L (the Fourier transform of the decay).

2.17.4. Gaussian Decay

Another useful function is an exponential decay that depends on the square of the time, called a *Gaussian decay*. The imaginary part of the continuous Fourier transform of a Gaussian decay cannot be expressed in closed form; however the real part is given by

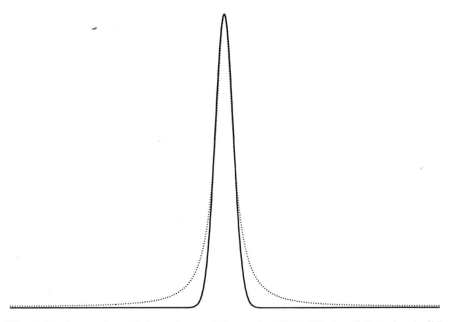

Fig. 2.11. Comparison of the real part of Lorentzian (dotted line) and Gaussian (solid line) peaks having the same FWHM shows that the Gaussian line shape decays to zero much more rapidly than the Lorentzian.

$$\text{real}[f(w)] = \text{real}\left(\int_0^\infty e^{-Gt^2}e^{-2\pi iwt}\,dt\right) = \sqrt{\frac{\pi}{4G}}\,e^{-(\pi^2/G)w^2} \qquad (2.47)$$

The half-maximum value is attained when $(\pi^2/G)w^2 = \ln(2)$, so the FWHM is equal to $(2/\pi)\sqrt{G\ln(2)}$. Equivalently, a Gaussian with FWHM equal to L will have $G = (\pi L/2)^2/\ln(2)$.

The Gaussian has an interesting property: Its Fourier transform is also a Gaussian, falling off as the exponential of the square of the frequency. Compared with the Lorentzian line shape, the Gaussian line shape falls off more rapidly. This can be seen in Figure 2.11, which shows Lorentzian and Gaussian line shapes that have identical FWHM. We'll see in the next chapter that converting Lorentzian line shapes into Gaussian line shapes is a useful way to improve resolution.

TO READ FURTHER

Books on the Fourier transform in both its continuous and discrete guises abound. Two that we have found particularly useful are by Bracewell [6] and Brigham [3]. The book by Bracewell covers both the continuous and discrete Fourier transforms. The book by Brigham is devoted to the discrete Fourier transform, and includes detailed discussions of various fast Fourier transform algorithms.

3

USING THE DFT: APPLICATION TO NMR

The fundamental aspects of the DFT present a majestic and orderly edifice; not so the real-life application of the DFT to NMR spectroscopy. Real NMR instruments suffer from more imperfections than merely the failure of the DFT to match the properties of the continuous Fourier transform. Much of the current practice of NMR data processing has been developed to alleviate these imperfections. To add a bit more excitement, different naming and sign conventions sometimes prevail, so a discussion of practical aspects of the DFT in NMR often resembles a boisterous, unmoderated debate. This chapter represents our attempt to moderate the proceedings. While the organization of the topics is not completely without logic, we must confess up front that most of what might be perceived as order is actually *trompe l'oeil*. Instead of approaching this material in a linear, straight-ahead fashion, feel free to sample as if it were a kind of antipasto. Check the subject headings to see what's in each dish.

3.1. NMR DATA SPAN AUDIO FREQUENCIES, NOT RADIO FREQUENCIES

NMR-active nuclei resonate at frequencies in the tens or hundreds of megahertz at magnetic field strengths in excess of 10 Tesla (characteristic of modern spectrometers). Yet the data from NMR experiments contain frequencies in the range of tens of kilohertz and lower. You paid big bucks to get a spectrometer that operates at as many megahertz as possible; have you been cheated? The answer is no. For each type of nucleus, the magnetic resonance frequencies span a range that is much smaller than the absolute frequency. For example, the hydrogen nuclei in a typical protein resonate in a frequency band that is less than 10 KHz wide centered at 500 MHz (in a magnetic field of 11.7 Tesla). Even if you could directly detect nuclear resonances at their Larmor frequencies—which for protons would require digitization rates in the hundreds of megahertz—the vast majority of the spectrum would be nothing but noise, except for the narrow band of frequencies containing the nuclear resonances. What the spectrometer does instead is to mix the signal from the sample with a signal that has the same frequency as the exciting radiation: the carrier frequency. The result is a signal that contains both the sum and difference of the carrier frequency with frequencies from the sample. The sum frequencies are very high and easily filtered out. The difference frequencies, which are recorded as the FID, fall in a narrow band centered at zero, typically spanning the audio range. Many modern computer workstations have sound generators that allow you to "listen" to NMR data; you would be surprised how rhythmic and oddly musical data from multidimensional experiments can sound.

3.2. ALIASING UNMASKED

A consequence of detecting NMR signals as the difference from the carrier frequency is that correct placement of the carrier is just as important as using a high sampling rate; improper choice of either can result in aliasing. Typically the carrier frequency is chosen to lie in the center of the range of frequencies emitted by the sample. The spectral width is then chosen to be large enough to include the entire range, so that none of the signals are aliased. However, by its very nature, aliasing is difficult to detect from a single experiment. One technique for determining if a peak is aliased is to acquire a second spectrum using a different spectral width. Non-aliased peaks will have the same frequency in both spectra. Peaks that have a frequency that is greater than the Nyquist frequency, but less that twice the Nyquist frequency ("aliased once"), will shift in frequency by ΔSW (the difference in the spectral widths). Peaks with frequencies greater than twice the Nyquist frequency, but less than three times the Nyquist frequency ("aliased twice"), will shift by 2ΔSW, and so on.

3.3. SPECTRUM DISPLAY ORDER

As we saw in Chapter 2, the frequencies of the DFT spectrum can be viewed in several different ways. The convention in NMR is to regard the frequencies as ranging from $-1/2\Delta t$ to $(N/2 - 1)/N\Delta t$ (with zero roughly in the middle). Since the carrier is typically placed in the center of the spectrum, this means that the frequencies of the DFT spectrum are equal to the frequency differences from the carrier of the actual signal. Another advantage is that the range of frequencies is nearly symmetrical around zero (the symmetry isn't perfect— you can't have zero exactly at the center if there is an even number of points!). Instead of the order $n = 0, \ldots, N - 1$, the elements of $\mathbf{f}$ from $n = N/2$ to

$N - 1$ are displayed first, followed by the elements from $n = 0$ to $(N/2) -$ 1, which simply requires swapping the two halves of the vector **f**.

But that is only part of the story. It is also the convention in NMR to display high frequencies to the left of low frequencies, so spectra are actually displayed with the halves swapped and then the entire spectrum reversed, taking care not to touch the zero-frequency component. The convention is illustrated in Figure 3.1. We hesitate to use the term "universal convention," since our experience

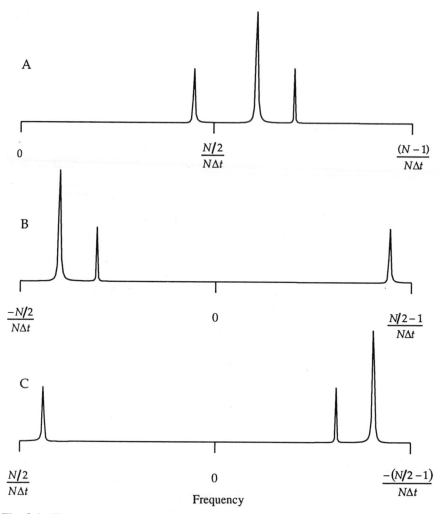

Fig. 3.1. The spectrum order corresponding to the definition of the DFT given by Eq. (2.1) has zero frequency at the left and the highest frequency on the right (**A**). Another common ordering convention places zero roughly in the center, negative frequencies on the left, and positive frequencies on the right (**B**). The order used in NMR puts negative frequencies on the right (upfield) and positive frequencies on the left (downfield, (**C**)).

is limited to this planet, and a small fraction of it at that.* Nevertheless, this ordering convention does appear to be widespread, a relic from the days of field-swept NMR.

A common misconception is that DFT spectra span the frequencies between $\pm 1/2\Delta t$. Not so! As we just saw, the upper limit is $(N/2 - 1)/N\Delta t$, which is one point below $1/2\Delta t$. Unfortunately, this misconception has been promulgated by some widely distributed NMR software. Check the software you are using: if the left- and right-most frequencies of your spectra are the same except for the sign, you've been victimized. Complain to your software provider. The error is not in the way these packages compute the DFT, but simply one of mislabeling the frequencies. If N is large, the error is small and probably not important. On the other hand, if N is small the error can be significant.

3.4. RELATIVE FREQUENCY SCALES AND THE CHEMICAL SHIFT

While the natural units for the frequency axis of DFT spectra are hertz or the index number n, these are not very convenient in NMR, since the frequency of NMR resonances depends on the magnetic field strength. It is more convenient to use a relative frequency scale, normalized to remove the dependence on field strength. The most widely used scale in NMR, the δ chemical shift scale, is defined by

$$\delta = \frac{w - w_{ref}}{w_0} \times 10^6 \tag{3.1}$$

where δ is the chemical shift of a peak with frequency w, w_{ref} is the frequency of a reference peak, and w_0 is the spectrometer carrier frequency. The convention of plotting high frequencies to the left of low frequencies still holds, so large chemical shift values appear to the left of small values. Although chemical shifts are dimensionless, they are described as having units of parts per million (ppm) to reflect the factor of 10^6. As an example, the proton spectrum of a typical protein contains peaks with frequencies ranging from $+10$ to -1 ppm, with respect to the reference frequency of the methyl protons in tetramethylsilane (TMS).

3.5. COUNTING DATA POINTS

Another widespread convention in NMR that can be confusing to the uninitiated is the practice of referring to the size of a data set in terms of the number of "real" points or the number of "complex" points. The use of the term "real"

*However, we are reminded of the abridged version of the Oxford English Dictionary, first published under the title "Oxford Universal Dictionary" in 1933. Who knows? Perhaps the notion was that the unabridged version would be too much for extraterrestrials.

does not imply that the data are real-only; instead it refers to the total number of component values. The term "complex" *does* imply that the data are complex, and it refers to the number of real-imaginary pairs. Thus a complex data set containing M complex points is also said to have $2M$ real points. (Of course, a real data set containing M values is said to have M real points.) To become a successful NMR spectroscopist it is important not only to master these conventions, but to keep a straight face while doing so.

3.6. CHARACTERISTICS OF NMR SIGNALS

A typical FID can be described as a sum of exponentially decaying sinusoids plus noise:

$$ d_k = \sum_{j=1}^{L} (A_j e^{i\phi_j}) e^{-k\Delta t/\tau_j} e^{2\pi i k\Delta t w_j} + \varepsilon_k \qquad (3.2) $$

where L is the number of sinusoids, A_j, ϕ_j, τ_j, and w_j are the amplitude, phase in radians, decay time*, and frequency, respectively, of the jth sinusoid, and ε_k is the random noise. Each sinusoid corresponds to a single nuclear resonance. From our consideration of exponential decay in the previous chapter, we know that such a signal will give rise to a spectrum that is the sum of Lorentzian peaks centered at the frequencies w_j. In reality, the decay of NMR signals is not always exponential. Furthermore, there are always experimental artifacts and other sources of nonrandom noise. Nevertheless, Eq. (3.2) is quite useful as a theoretical model.

This modeling can take the form of simulation, using given values of the parameters w_j, τ_j, and so on. Alternatively, modeling can take the form of parametric spectrum analysis, in which the parameters are fit to match experimental data. We discuss parametric spectrum analysis in Chapters 4 and 6. Spectrum simulation as a tool for error analysis is discussed in Chapter 7.

3.7. ZERO-FILLING AS EXTRAPOLATION

The noise term ε_k in Eq. (3.2) remains more or less constant over time, whereas the sinusoidal parts inexorably decay. Eventually there comes a time when the noise contributes more to the FID than do the nuclear resonances. At that point it makes sense to stop sampling, since additional data would mainly add noise. Even if the signal has not decayed below the noise level, there often is a

---

*In NMR experiments, the decay time is called T_2. Sometimes you will see T_2^* instead. T_2^* is an effective decay time that includes the influence of magnetic field inhomogeneity. For our purposes we can neglect the inhomogeneity. Don't you wish you could when you are shimming the spectrometer?

practical limit to the number of samples that can be recorded. In order to obtain high digital resolution with the DFT, however, it is necessary to have a lot of samples: the larger N, the smaller the frequency spacing $1/N\Delta t$ of the DFT spectrum. Instead of collecting N samples, we can collect a smaller number M, and then try to extrapolate the FID (somehow) out to N points. If the signal has decayed, a good extrapolation is simply to add zeros, a process called *zero-filling*. Zero-filling is equivalent to a kind of interpolation in the frequency domain; at the frequencies they have in common, the DFT spectra of the zero-filled and the unextended data will agree (up to a scale factor).

A zero-filled FID is equal to a fully sampled N-point FID that has been multiplied by a step function, with the step located at M (Fig. 3.2B). According to the convolution theorem, the resulting spectrum is equal to the spectrum of the fully sampled FID convolved with the spectrum of the step function, with all its attendant sinc wiggles. A comparison of the zero-filled DFT with the DFT of the original, unextended data is shown in Figure 3.2C. The filled circles in the figure correspond to frequencies that the two spectra have in common; the line connects the points of the spectrum of the zero-filled FID.

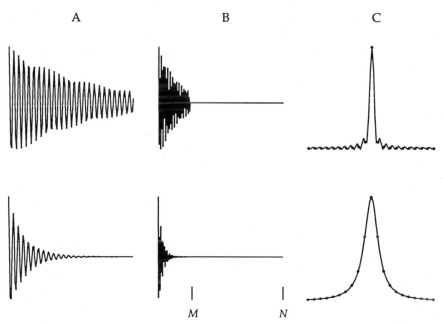

A   **B**   **C**

M N

Fig. 3.2. Zero-filling, a form of extrapolation in the time domain, can be viewed as a form of interpolation in the frequency domain. The signals shown in (**A**) have been zero-filled from $M = 128$ to $N = 512$ points (and compressed for display purposes) in (**B**). The DFT spectra of the zero-filled signals are displayed in (**C**). The filled circles indicate the points making up the spectra of the nonextrapolated data. Clearly, the interpolation corresponding to zero-filling is a more reasonable approximation for signals that have decayed to near zero.

Whether or not this kind of interpolation makes sense depends on whether the signal has decayed by the end of the sampling interval. If it has (bottom row of Fig. 3.2), then zero-filling makes good sense. If it hasn't (top row), the leakage resulting from zero-filling will cause severe problems. In fact, the interpolation shown in the top row can fairly be described as wacko.

How much should one zero-fill? Zero-filling by at least a factor of two enforces causality. Beyond that, the answer depends on how much the signal has decayed by the end of the sampled interval and on the desired digital resolution for the final spectrum.

3.8. MINIMIZING LEAKAGE: APODIZATION

A powerful tool for minimizing the leakage that comes from zero-filling is *apodization*, also called *windowing* or *data tapering*. Apodization is the process of modifying the data—prior to zero-filling and Fourier transformation—in a way intended to improve the quality of the spectrum. This may sound a bit vague: There are many different measures of quality, and different apodization functions can be used to improve DFT spectra in different ways, although we shall see that it is not possible to improve all measures of spectral quality at once.

Apodization is performed by multiplying the data by a vector **a**, element by element:

$$d'_k = a_k d_k \tag{3.3}$$

The apodization function **a** is real, so that apodizing complex-valued data leaves the phase and frequency unchanged. The envelope of the data *is* changed, however. (Broadly speaking, the *envelope* can be described as a smooth curve that just touches the highest points of the FID.) Apodization functions used for leakage reduction go smoothly to zero at the end of the sampled interval, which minimizes the extent of the discontinuity when the modified data vector **d**' is zero-filled (Fig. 3.3). We will give some examples of apodization functions later. Note that Eq. (3.3) implies that apodization is linear, which means that apodization can be used in applications requiring quantitative analysis.

3.9. APODIZATION FOR SENSITIVITY OR RESOLUTION ENHANCEMENT

Apodization can be used to enhance spectra in ways other than by reducing leakage. There are window functions designed to improve the sensitivity of a spectrum, by weighting more heavily the points where the signal is large compared to the noise. Window functions can also be designed to improve resolution, by modifying the decay envelope to obtain narrower lines. Unfortu-

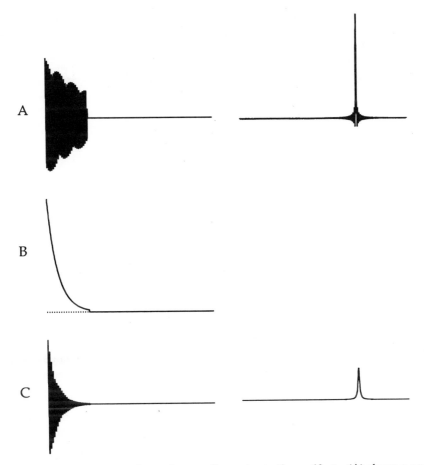

Fig. 3.3. Apodization can be used to ameliorate truncation artifacts. (**A**) shows a zero-filled signal and its DFT spectrum, complete with sinc wiggles. Apodization using the function in (**B**) (an exponential decay) yields the data in (**C**). The sinc wiggles in the spectrum have been reduced, but at the expense of broadening the peak.

nately, it is difficult to find a window function that does both at the same time. To see why, consider that the way to improve resolution is to use an increasing window function that partially cancels the decay. Such a function will also amplify the noise near the end of the FID, resulting in reduced S/N (Fig. 3.4A and B). Instead of a growing apodization function, a decaying function can be used to improve S/N (Fig. 3.4C). Of course the gain comes at the expense of resolution, since the function hastens the decay of the signal envelope. This trade-off between sensitivity and resolution is characteristic of linear filtering methods. We will see in Chapter 5 that nonlinear methods can sometimes achieve gains in resolution with no loss in S/N.

An important concept in signal processing is that of the *matched filter*, which

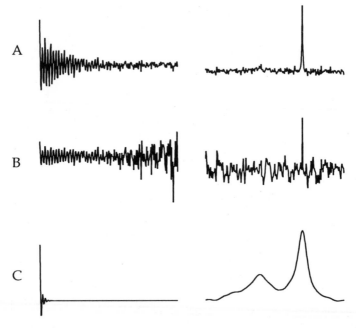

Fig. 3.4. Apodization can be used to improve resolution or sensitivity, but not both simultaneously. **(A)** shows an unapodized signal and its DFT spectrum. In **(B)**, apodization by an increasing exponential has improved the resolution (the peak is narrower) but decreased the S/N. In **(C)**, apodization by a decreasing exponential has improved the sensitivity (both peaks are now visible), but at the expense of greatly broadening the peaks.

is an apodization function that matches the envelope of the signal. It can be shown by a statistical argument that the matched filter best maximizes the S/N of the resulting spectrum [7]. However, the matched filter also results in a loss of resolution, by broadening spectral components. For a Lorentzian line the envelope decays as $e^{-\pi Lt}$, so multiplying by its matched filter,

$$(e^{2\pi iwt}e^{-\pi Lt})e^{-\pi Lt} = e^{2\pi iwt}e^{-2\pi Lt} \tag{3.4}$$

results in a Lorentzian line shape that is broadened by a factor of two. Similarly, for Gaussian lines, applying the matched filter,

$$(e^{2\pi iwt}e^{-Gt^2})e^{-Gt^2} = e^{2\pi iwt}e^{-2Gt^2} \tag{3.5}$$

results in broadening by a factor of $\sqrt{2}$, since the line width is proportional to $\sqrt{G}$.

While apodization is valuable for improving DFT spectra, it is not without limitations. In addition to the fact that it can't improve resolution and sensitivity simultaneously, apodization affects *all* components indiscriminately, regardless of their decay rates or frequencies. A window function that is a matched filter for one peak will not be a matched filter for peaks that have different widths. The same window function can have qualitatively different effects on sinusoids with different decay rates. The right side of Figure 3.4C shows a spectrum obtained using a matched filter for the broad peak. Notice that the narrow peak is broadened, but still visible, as it was without apodization. The broad peak, on the other hand, stands out much more distinctly than it does in the spectrum of the nonapodized data (Fig. 3.4A) or the spectrum of the data apodized to improve resolution (Fig. 3.4B).

3.10. A CHILD'S MENAGERIE OF APODIZATION FUNCTIONS

The large number of different apodization functions in the scientific and technical literature has sometimes been described as a zoo. Here is a taxonomy of some of the more useful species for NMR, including their common names. Examples of each are shown in Figure 3.5.

3.10.1. Exponential Apodization

The exponential apodization function (common name EM, for exponential multiplication), which we've already seen, is given by

$$a_k = e^{-\pi W k \Delta t} \tag{3.6}$$

where the broadening constant W, in hertz, can be positive (for sensitivity enhancement) or negative (for resolution enhancement). EM is important because it preserves the Lorentzian nature of lines, and because it often gives a good approximation to the matched filter. A Lorentzian line having width L prior to apodization will have width $L + W$ following apodization by EM. One limitation is that if the broadening factor W is not sufficiently large, EM does not go close enough to zero to prevent leakage. Also, it is not a good approximation to the matched filter for sine-modulated signals, such as those that arise from COSY (correlation spectroscopy) experiments.

3.10.2. Gaussian Apodization

Gaussian apodization (GM, for Gaussian multiplication) is similar to EM, except that the exponential decay depends on the square of the time:

$$a_k = e^{-W(k \Delta t)^2} \tag{3.7}$$

Real line shapes are often intermediate between Lorentzian and Gaussian, and sometimes are more nearly Gaussian than Lorentzian. For these signals, GM is closer to the matched filter. As with EM, leakage can be a problem if W is

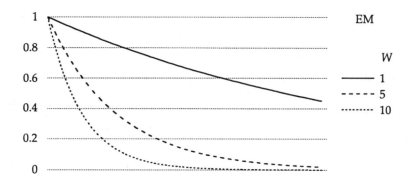

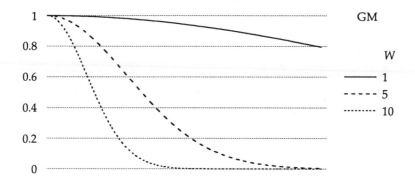

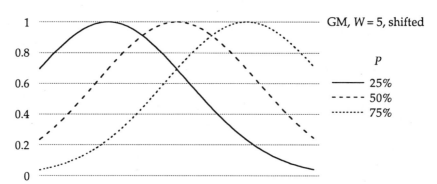

Fig. 3.5. An apodization menagerie. For each type of function, several curves are plotted showing the effects of different choices for the adjustable parameters.

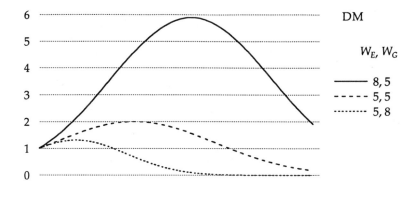

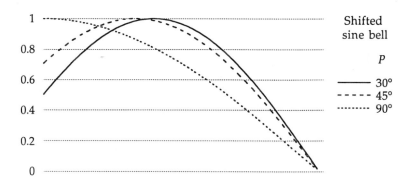

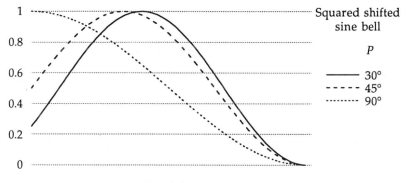

Fig. 3.5. (*Continued*)

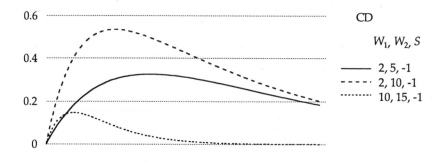

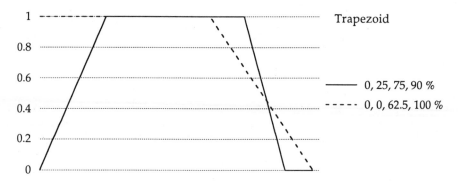

Fig. 3.5. (*Continued*)

not sufficiently large for the apodized signal to approach zero at the end of the sampled interval.

3.10.3. Shifted Gaussian Apodization

EM and GM correspond to matched filters only for signals whose envelopes are maximum at time $t = 0$. In many NMR experiments the maximum of the envelope is not at time zero. One way to deal with data that has a maximum later than time zero is to shift the entire curve so that the maximum of the apodization function falls somewhere in the middle of the sampled interval. Shifting an exponential decay doesn't accomplish anything, since the shape of the curve will remain unchanged. A Gaussian, on the other hand, has a definite maximum at $k = 0$, which can be shifted by an arbitrary amount:

$$a_k = e^{-W(k\Delta t - PM\Delta t)^2} \tag{3.8}$$

where P is the amount of the shift expressed as a fraction of the total data length M.

3.10.4. Lorentz-to-Gauss Transformation

Lorentz-to-Gauss transformation (DM, for double exponential multiplication) is often a useful apodization function for lines that are nearly Lorentzian. The underlying principle is to convert Lorentzian lines into Gaussian lines by multiplying by an increasing exponential (to cancel the decay) followed by a decreasing Gaussian (to introduce a Gaussian decay). This is the same as applying a Lorentzian narrowing and a Gaussian broadening (a negative EM followed by a positive GM):

$$a_k = e^{+\pi W_E k \Delta t} e^{-W_G (k\Delta t)^2} \qquad (3.9)$$

where W_E is the exponential narrowing constant and W_G is the Gaussian broadening constant. A virtue of DM is that the maximum can be shifted by appropriate choice of the constants. Like EM and GM, however, it may not approach zero at the end of the sampled interval. In fact, it can grow very large if W_G is too small in comparison to W_E.

3.10.5. Sine Bell Apodization

A versatile apodization function that goes smoothly to zero and can be adjusted so that its maximum falls in the middle of the sampled interval is sine bell:

$$a_k = \sin \left(\frac{180° - P}{M} k + P \right) \qquad (3.10)$$

where P is between zero and 90°. The apodization function starts from the value of the sine function at the phase shift P and continues to 180° at the end of the sampling period. The phase shift provides great flexibility, yet the number of parameters is small. Depending on P, the sine bell can be used to enhance sensitivity or resolution. A sine bell shifted by 90° is called a cosine bell, for the obvious reason. There is also a squared sine bell function:

$$a_k = \sin^2 \left(\frac{180° - P}{M} k + P \right) \qquad (3.11)$$

It has many of the properties of the sine bell, but it is more concentrated around the maximum.

3.10.6. Convolution Difference

Another versatile apodization function, but one that is more complicated (it has three adjustable parameters) is convolution difference (CD):

$$a_k = e^{-\pi W_1 k \Delta t} - S e^{-\pi W_2 k \Delta t} \qquad (3.12)$$

where W_1 and W_2 are width parameters and S is a dimensionless scale factor. CD apodization is simply the weighted difference between the results of apodizing with EM using two different broadening factors. Characteristic negative wings appear when using $S = 1$; they can be minimized by using a smaller value. While CD is not widely implemented, it can be performed using any software that provides EM and allows the computation of difference spectra, which virtually all programs do.

3.10.7. Trapezoid Apodization

A trapezoid is a primitive but still very useful apodization function. Most NMR software packages provide a means of specifying a trapezoid in terms of the positions of a rising edge (going linearly from zero to one) and a falling edge (going from one to zero), with the function taking the value one between the edges and zero elsewhere. The four parameters needed to specify the edges can take the form of indices or percentages of the total length of the data. For example, a trapezoid specified by the indices 0, 64, 128, 256 for a 256-point FID can also be specified by the percentages 0.0, 25.0, 50.0, 100.0. Although the sharp corners at the edges of the trapezoid can lead to sinc wiggles, judicious choice of their location or the use of a smooth apodization function (in conjunction with the trapezoid) can minimize artifacts.

3.11. APODIZATION: RULES OF THUMB

Just as the proof of the pudding is in the eating, so too the best choice of a window function is determined by the results. Since equally good results can be obtained using different window functions, the most convenient window function is largely a matter of personal taste. Most modern NMR data processing programs can simultaneously display the effect of a window function on the FID and on the spectrum, while allowing the user to interactively adjust the parameters. This makes it easy to become adept at optimizing window functions. As Figure 3.6 shows, the choice of parameters can have significant consequences.

Three useful rules of thumb are the following: To enhance sensitivity, use an apodization function that is close to the matched filter. For resolution enhancement, use one that emphasizes later times compared to early times (because the slower the decay, the narrower the lines). To suppress leakage, use a function that smoothes out discontinuities and sharp corners. The only way to find out what apodization works best for your data is to experiment. Unfortunately, the expression "cooking your data" has taken on a pejorative meaning, but the analogy may be more apt and desirable than the derogatory usage suggests. Good cooks rely on the best available ingredients, and use seasonings judiciously to bring out the full potential of the flavors locked inside. A good

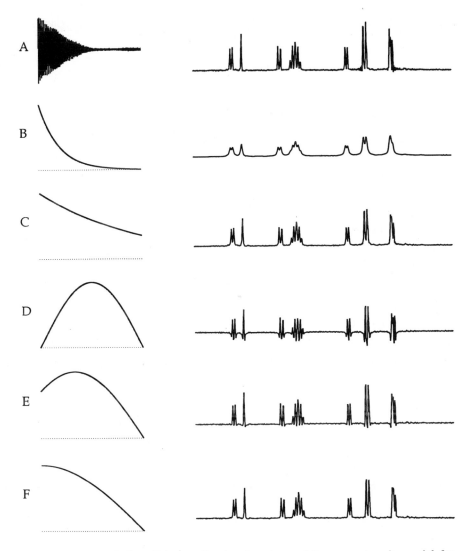

Fig. 3.6. Proper choice of the apodization function and its parameters is crucial for obtaining the highest quality spectrum. (**A**) shows a signal and its DFT spectrum. In (**B–F**), various apodization functions are shown together with the DFT spectra of the apodized data. The functions are (**B**) EM 5.0, (**C**) EM 1.0, (**D**) sine bell, (**E**) sine bell shifted by 45°, and (**F**) sine bell shifted by 90°. The original spectrum (**A**) contains sinc wiggles at the base of the large peaks; they are gone in (**B**), but the resolution is considerably poorer. (**C**) demonstrates a good compromise. (**D**) shows the negative wings that are characteristic of overaggressive resolution enhancement; they are somewhat reduced in (**E**), but still visible. (**F**) is very similar to (**C**), despite the different shapes of the apodization functions.

spectroscopist will seek out the highest quality data, and apodize it judiciously to bring out the information it contains.

3.12. DATA ACQUISITION SCHEMES

The choice of the apodization function is dictated by the characteristics of the FID. In addition to these intrinsic characteristics, however, there are many aspects of the NMR experiment that influence the way the data must be processed. These include all of the details of the acquisition scheme: whether the data are real or complex, whether the sampling is linear or nonlinear, whether the data are oversampled, and to what extent the start of data acquisition is delayed. (If some of these topics are unfamiliar, don't worry; we will discuss each of them later on.)

3.12.1. Time-Proportional Phase Incrementation (TPPI)

We saw in Chapter 2 that data can be either real or complex. With proper care, it is possible to collect real-only data in a way that affords many of the advantages of simultaneous quadrature detection. Here's how to do it: Since single-channel data does not distinguish between positive and negative frequencies (lack of sign discrimination), every peak will show up twice in the DFT spectrum—on the positive side and on the negative side. If we can arrange things so that all the true signals have negative frequency, for example, we can throw away the positive half of the spectrum, and all will be well. That means doubling the spectral width, since we only keep half of the spectrum. One way to get all of the signals to have negative frequency is to place the carrier at a frequency just higher than all the true signals. Alternatively, the carrier can be centered among the signals, and the spectrum can be shifted by clever manipulation of the detector phase, called *time-proportional phase incrementation*, or TPPI.

Suppose that after each sample point the detector phase is incremented by $90°$. With a single detector we then measure the following sequence of components of the signal:

$$
\begin{array}{ccccccccc}
k & 0 & 1 & 2 & 3 & 4 & 5 & 6 & \ldots \\
\text{component} & X & Y & -X & -Y & X & Y & -X & \ldots
\end{array}
\tag{3.13}
$$

instead of all X components. The result is the same as if we multiplied the original complex signal by the sequence

$$
\mathbf{a} = 1 \ -i \ -1 \ i \ 1 \ -i \ldots
\tag{3.14}
$$

and then just detected the X component. The sequence (3.14) can be expressed as

$$a_k = e^{-\pi i k / 2} \qquad (3.15)$$

We saw in Chapter 2 that demodulation by $e^{-2\pi i w t}$ has the effect of shifting frequencies by $-w$; in this case the effect of incrementing the phase is to shift the spectrum by $-1/4\Delta t$. This corresponds to one quarter of the doubled spectral width (or one half of the original spectral width), so the peaks are all moved to the negative half of the spectrum. Fourier transformation of the real-only data yields a spectrum that is symmetric about zero frequency (Fig. 3.7A). Discarding the positive half of the spectrum and shifting the frequency axis so that the origin coincides with the carrier frequency then results in a spectrum nearly identical to what would be obtained using simultaneous quadrature detection (Fig. 3.7B).

Nearly identical, but not exactly. TPPI differs from simultaneous quadrature

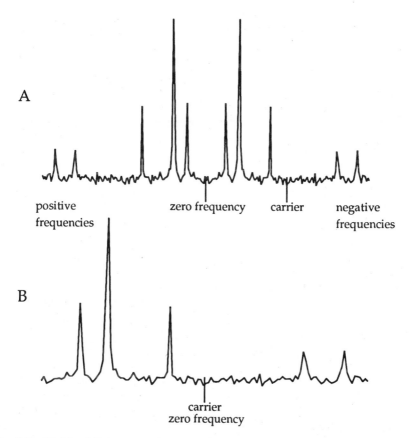

Fig. 3.7. (A) The full DFT spectrum of TPPI data is symmetrical because the data are real-only. The effect of the time-proportional phase incrementation is to shift zero frequency away from the carrier frequency. (B) Discarding the left half of the spectrum and shifting the frequency origin yields nearly the same results as quadrature detection.

detection in a couple of respects. One is that samples are collected twice as frequently (as a result of doubling the spectral width). But since each sample is only a real, rather than a complex number, the total number of data values is the same. Another difference is the way that aliasing shows up. With simultaneous quadrature, peaks just beyond the positive end of the spectrum appear to be wrapped around to just inside the negative edge. With TPPI, such peaks appear to be reflected back inside the positive edge of the spectrum. This is not a violation of the periodicity of the DFT; what you are really seeing is a result of the lack of sign discrimination in the double-width spectrum.

Suppose that your spectrometer is only capable of shifting the detector phase by 0° or 90°. No sweat. Just alternate the phase for successive samples, collecting the components

$$X \quad Y \quad X \quad Y \quad X \quad Y \quad X \quad Y \qquad (3.16)$$

and change the sign of alternate pairs of points to arrive at the original TPPI sequence (3.13). This is known as the Redfield trick* [8].

There is hardly an NMR spectrometer in use today that is not capable of simultaneous quadrature detection, or shifting phase by nearly arbitrary amounts, for that matter. Since simultaneous quadrature detection is less demanding in terms of sampling rate, you might reasonably wonder whether TPPI has anything other than historical interest. In fact, it is alive and well and in common use. We'll see why when we discuss multidimensional NMR data processing.

3.12.2. Oversampling

The sampling theorem tells us that sampling at less than the Nyquist frequency results in aliasing. But what happens if we sample faster—several times faster? At first glance it might appear that the additional effort would be wasted, since it would extend the spectral width beyond the range of frequencies contained in the signal, and the extra bandwidth will only contain noise. It turns out, however, that the extra sampling can be worthwhile. Consider an FID over-sampled by a factor of four, so the sampling rate is $4W$ rather than W (i.e., the sample d_k is collected at time $k/4W$). The samples d_0, d_4, d_8, . . . are the samples that would have been collected by sampling at the original rate W. The samples d_1, d_5, d_9, . . . form another data set sampled at the original rate; the full oversampled FID $\mathbf{d}$ can be decomposed into four such sub-FIDs. We can select out these sub-FIDs by multiplying $\mathbf{d}$ by appropriate mask functions that take the value one at every fourth point and are zero elsewhere. The Fourier transform of each masked FID is just the convolution of $\mathbf{f}$ (the DFT of $\mathbf{d}$) with the DFT of the mask function; the four mask functions and their DFTs are shown in Figure 3.8.

---

*Is it a coincidence that two unrelated scientists, Richard Ernst and Al Redfield, responsible for so many fundamental theoretical and experimental contributions to modern NMR spectroscopy, are also wizards at signal processing? We don't think so.

A   B

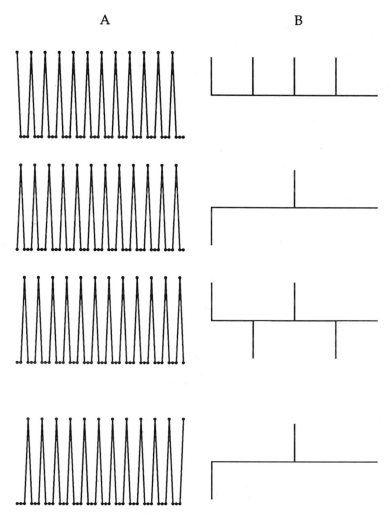

Fig. 3.8. (**A**) contains the four possible masking functions in which every fourth point is one and the other points are zero. (**B**) contains the corresponding DFT spectra. Notice that the zero frequency component has the same phase in each spectrum, whereas the other components cycle in phase.

The spectrum of the first mask function consists of four positive peaks, at frequencies $-2W$, $-W$, 0, and W (corresponding to $n = N/2$, $3N/4$, 0, and $N/4$). The second mask function is the same as the first, but shifted in time by one sample period; thus the DFT spectrum is the same as for the first, but with a linear phase shift of 360°. The peak at zero frequency remains unaffected, but the other peaks experience phase shifts ranging from 90° to 270°. The remaining mask functions correspond to additional time shifts, and their DFT spectra experience an additional 360° linear phase shift with each time shift. The effect of convolving the spectrum of a mask function with $\mathbf{f}$ is to generate four copies of **f**, shifted in frequency by $-2W$, $-W$, 0, and W. These copies are also shifted in phase, according to the phase of the corresponding peaks in the mask spectrum (Fig. 3.9).

The sum of the masked FIDs is equal to **d**, so the sum of the spectra of the masked FIDs is equal to **f**. The peaks shifted in frequency by mW go through m full cycles of phase, and when added together they exactly cancel unless $m = 0$. The peaks shifted by zero, on the other hand, add constructively, so that the sum has four times the intensity of the peaks in the masked spectra. This

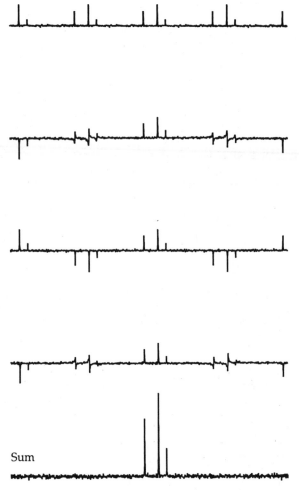

Fig. 3.9. The first four rows contain the DFT spectra of an oversampled signal that has been masked using the four functions in Figure 3.8. Each peak appears four times, with varying phases. The bottom row shows the sum, which is the same as the DFT spectrum of the full oversampled signal. Notice that the signal peaks add constructively only in the center band, while the noise remains distributed over the full spectral width.

is the same as saying that oversampling by a factor of four leads to a factor of four increase in the intensity of the peaks. If the noise in the four masked spectra is uncorrelated, then it adds up only as the square root, in this case two. Thus the S/N improves by a factor of two, as can be seen in the bottom row of Figure 3.9.

A simple way of thinking about the gain in S/N is that oversampling allows us to detect and remove noise that would otherwise be aliased into the spectrum. The examples shown in Figure 3.9 included synthetic noise from a random number generator, which is not band-limited. On real spectrometers, however, some of the noise sources *are* band-limited. The analog circuitry includes a band-pass filter, which rejects high-frequency noise components in the signal from the probe. Therefore, noise from sources occurring before the filter will be affected by oversampling only to the extent that the filter passband exceeds the original spectral width. Noise from sources occuring later in the signal path will benefit more from oversampling (if they are not also band-limited). Beckman and Zuiderweg have undertaken a detailed empirical analysis of the benefits and limitations of oversampling in NMR, including simple tests for determining what type of noise dominates an experiment [9].

There are other benefits of oversampling, however. One is that the effective *dynamic range* of the spectrum is improved. (The dynamic range is the ratio between the largest and smallest values.) The analog-to-digital (A/D) converters in the spectrometer discretize the value of the measured signal, and the precision of the measurement is limited by the digital resolution of the A/D converter. This resolution depends on the length of the digital word, or number of binary digits (*bits*) used to represent the result of a measurement. Modern high resolution NMR spectrometers typically have A/D converters that produce a 16-bit number, which can represent values in the range -2^{15} to $2^{15} - 1$, or $-32{,}768$ to $+32{,}767$. The dynamic range between the highest and lowest numbers is often measured in decibels (dB). A decibel is defined as 20 times the $\log_{10}$ of the range; a 16-bit A/D converter has a dynamic range of $20 \cdot \log_{10}(32{,}768) = 96$ dB, while a 12-bit converter has a dynamic range of 72 dB. The coherent addition of subspectra described above shows that oversampling by a factor of four allows us to obtain a spectrum equivalent to one without oversampling, but with dynamic range four times larger, corresponding to an additional 2 bits, or an additional 12 dB. In general, the increase in dynamic range that accompanies oversampling by a factor of K is $\log_2(K)$ bits, or $20 \cdot \log_{10}(K)$ dB.

The drawback of oversampling, of course, is that the amount of data collected goes up linearly with the amount of oversampling. There is a way to avoid the problem of overabundant data, while maintaining the benefits of oversampling. Using high-speed digital signal processors, it is possible to digitally filter and *decimate** a signal to a lower sampling rate. These processors

---

*Decimate literally means to keep every 10th point and throw away the rest; however, it is commonly used to describe data reduction by arbitrary amounts.

are widely used in consumer compact disk players, and are now available in commercial NMR spectrometers. Nonetheless, this procedure is not without drawbacks of its own. The digital filters employed in spectrometers are sometimes proprietary, so that you don't really know what has been done to the data. In addition, some of the subsequent processing steps may interact in unexpected ways with the digital filters. Given the continuing deflation in disk prices, our advice is to forego digital filtering and decimation in the time domain, and to collect the full oversampled FID. The size of the spectrum can be reduced later by throwing out the portions beyond the original spectral width. The additional computational cost is hardly a burden with modern computer workstations.

3.13. PHASE CORRECTION

When we introduced the Lorentzian, the absorption line shape was neatly confined to the real part of the spectrum and the dispersion line shape was confined to the imaginary part. In practice, one rarely sees such "pure phase" spectra following Fourier transformation. For a variety of reasons, the absorption and dispersion line shapes are mixed, resulting in line shapes (Fig. 3.10B and C) that lack the symmetry of pure absorption or pure dispersion line shapes (Fig. 3.10A and D). Another way to put it is that in the description of the FID as a sum of decaying sinusoids [Eq. (3.2)], the phases ϕ_j need not all be zero. This effect can arise from a difference between the phases of the spectrometer's transmitter and detector, or a delay in the start of acquisition. In many instances it is possible to unscramble the mixed signal to obtain pure phase line shapes; the process is called *phase correction*.

Provided that the mixture of absorption and dispersion is the same for all peaks in the spectrum, the pure phase spectrum $\mathbf{f}'$ can be obtained by means of a phase shift, or rotation of the elements of $\mathbf{f}$ in the complex plane:

$$f_n' = e^{i\varphi}f_n \tag{3.17}$$

Since the amount of rotation is constant, that is, independent of frequency, the phase shift can equally well be performed in the time domain:

$$d_k' = e^{i\varphi}d_k \tag{3.18}$$

where $\mathbf{d}'$ can be Fourier transformed to obtain $\mathbf{f}'$.

Often, frequency-independent phase correction isn't enough to unscramble the mixing of absorption and dispersion. For example, a shift in the time domain—such as a delay in the start of acquisition—results in a phase shift that varies linearly with frequency, as we saw in Chapter 2. The appropriate phase correction will have the form

$$f_n' = e^{i(\phi_0 + n\phi_1/N)}f_n \tag{3.19}$$

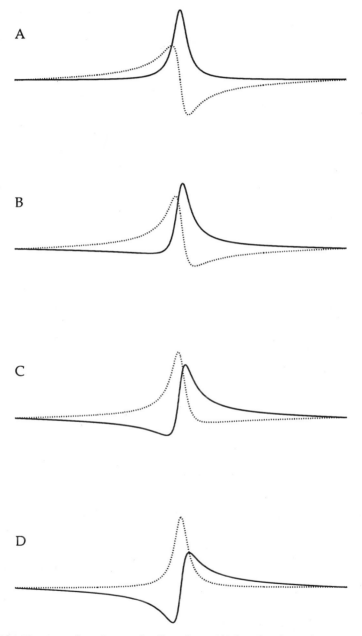

Fig. 3.10. The pure phase Lorentzian line shape (**A**) has the absorptive component in the real part (solid line) and the dispersive component in the imaginary part (dotted line). Lorentzians that are phase-shifted by 30° (**B**) and 60° (**C**) contain a mixture of absorptive and dispersive components. A 90° phase-shifted Lorentzian (**D**) preserves the separation between absorptive and dispersive components, but they are interchanged between the real and imaginary parts.

where ϕ_0 and ϕ_1 are the constant and linear components of the phase correction, respectively. [Note that in Eq. (3.19) ϕ_0 and ϕ_1 are in radians, but phase corrections are more often expressed in degrees.] The explicit dependence on frequency means that this mixing can only be performed in the frequency domain.

In principle it is possible to compute automatically the constant and linear phase corrections that result in pure phase spectra, and most data processing programs do a fairly good job. Experiments that give rise to both positive and

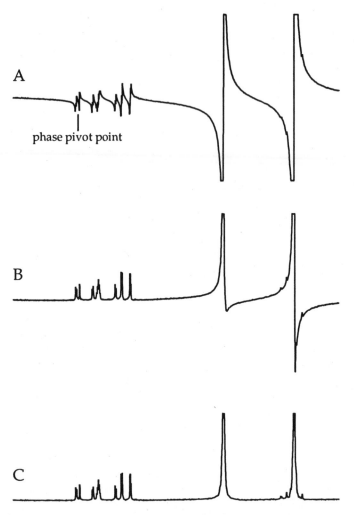

Fig. 3.11. Spectrum phasing can be performed in two steps using a phase pivot point. (**A**) shows an unphased spectrum. In (**B**), a constant phase correction has been applied to bring the peak at the pivot point into phase. In (**C**), a linear phase correction about the pivot point suffices to bring the remaining peaks into phase.

negative peaks, however, or that have very broad peaks, can be difficult to phase automatically. A practical strategy for manually determining the constant and linear phase corrections is to choose one peak as a "phase pivot point" (Fig. 3.11A). The peak at the pivot point is phased using only a constant correction (Fig. 3.11B). A linear correction is then applied to bring the rest of the spectrum into phase, using the selected point j as the pivot; that is, a combination of constant and linear phase shifts is used to phase the spectrum while keeping the phase of f_j unchanged (Fig. 3.11C). The overall phase correction is given by

$$f_n' = e^{i(\phi_0 + (n-j)\phi_1/N)}f_n \tag{3.20}$$

Phasing by ϕ_0 and ϕ_1 with a pivot point at j is equivalent to a phase correction by $\phi_0' = \phi_0 - j\phi_1/N$ and $\phi_1' = \phi_1$ with a pivot point at zero, the form given by Eq. (3.19). It's a lot easier to keep track of the overall phase shift in terms of this form, especially when phasing interactively and perhaps even moving the pivot point around.

If your sole purpose is to obtain spectra with only positive peaks, there are simpler methods: namely, computation of the *power* or *magnitude* spectra, $|f_n|^2$ or $|f_n|$, respectively. (Don't apply these procedures until after the baseline has been corrected; see below.) Neither depends on the phase of the original spectrum, and both result in positive spectra. Neither is appropriate, however, if sign discrimination or quantification is important (as in nuclear Overhauser effect spectra). Furthermore, peak intensities are distorted in power spectra, because of the nonlinear squaring operation, and magnitude spectra have peaks that are significantly broadened, compared to the pure phase line shape (Fig. 3.12).

3.14. QUADRATURE IMAGES

It should come as no suprise that if the two components of simultaneous quadrature data are not exactly 90° different in phase, "ghost" images of the peaks reflected about zero frequency appear. While these *quadrature images* (or *quad images* for short) can result from a variety of instrumental problems, it has been our experience that they are often the result of incorrect data processing. Either the data organization has been confused, the data have been treated as TPPI when they should have been treated as complex (or vice versa), or the Redfield trick has been neglected (or applied) inappropriately. Of course, there is also the possibility (gasp!) that the data were not collected properly to begin with. It's surprisingly easy to foul up the detection scheme for indirect dimensions of multidimensional experiments, even on the most modern spectrometers. These sorts of mistakes almost always show up as quad images that have intensity equal to that of the real peaks, or as reversed spectra.

Quad images that are weaker in intensity than the real peaks often result

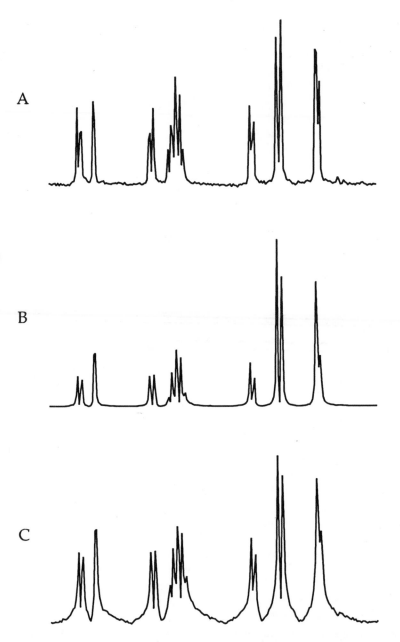

Fig. 3.12. Comparison with the real part of a properly phased DFT spectrum (**A**) shows that the power spectrum (**B**) has distorted peak intensities and the magnitude spectrum (**C**) has broadened peaks.

from more subtle problems with the spectrometer. There are two types of problems that can be corrected by data processing: if the receiver gain is different for the real and imaginary channels, or if the phase difference is not exactly 90°. When appropriate phase cycling and signal routing are used, these effects will be averaged out. If only one transient is collected, however, the imperfections will remain. The data can be repaired by the following procedure, which is nothing more than the familiar Gram-Schmidt process for constructing an orthonormal basis. The basic idea is to compute the correlation between the real and imaginary channels, then subtract the appropriately weighted correlation from one of the channels. Before starting, the DC (zero frequency) component must be removed from the real and imaginary parts of the data (see the next section). The correlation is given by the sum

$$C = \sum_{k=0}^{M-1} \text{real}(d_k)\text{imag}(d_k) \tag{3.21}$$

and the total power in the imaginary channel is

$$I = \sum_{k=0}^{M-1} [\text{imag}(d_k)]^2 \tag{3.22}$$

The real part of the data is modified to remove the correlation:

$$\text{real}(d_k') = \text{real}(d_k) - \frac{C}{I}\text{imag}(d_k) \tag{3.23}$$

and finally the power of the real part is adjusted to match the level of the imaginary part:

$$\text{real}(d_k'') = \frac{I}{S}\text{real}(d_k'), \quad \text{where} \quad S = \sum_{k=0}^{M-1} \text{real}(d_k')^2 \tag{3.24}$$

The imaginary part of $\mathbf{d}''$ is set equal to the imaginary part of $\mathbf{d}$. It is straightforward to show that the real and imaginary parts of $\mathbf{d}''$ are orthogonal, that is, the sum (3.21) computed for $\mathbf{d}''$ vanishes. Figure 3.13 shows quad images that result from unbalanced gain in the real and imaginary channels, and the spectrum of the corrected data.

Quad images due to channel imbalance were not uncommon in the early days of FT-NMR; many early software packages provided a "QC" or quadrature correction orthogonalization procedure. They have become less of a problem because of the common practice of alternating signal routing for multiple transients. The modern use of pulsed field gradients sometimes removes the need to cycle the phase and hence to average multiple transients; as a result, Gram-Schmidt orthogonalization may once again become common.

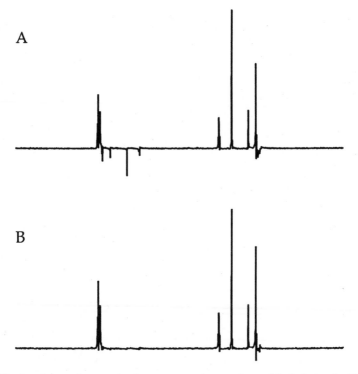

Fig. 3.13. Quadrature images in the spectrum due to channel imbalance (a difference in gain in the real and imaginary channels) (**A**) can be removed by Gram-Schmidt orthogonalization (**B**).

3.15. THE BATTLE AGAINST ARTIFACTS

Quadrature images and leakage are not the only artifacts that appear in NMR spectra. Artifacts are a little like weeds, and like a farmer, a spectroscopist needs to be both crafty and vigilant in order to prevent them from ruining the harvest. Here are some other common artifacts, and some tools for dealing with them.

3.15.1. Zero-Frequency Spike

One artifact that is especially easy to deal with is a spurious zero-frequency peak. A way to tell if a peak at zero frequency is spurious is to shift the carrier frequency slightly; if the peak remains at zero frequency, it's spurious. The cause and the solution are both simple. The zero-frequency component of the DFT spectrum is proportional to the average value of the FID; if this average is nonzero there will be a spike. The spike can be removed by adding a constant to the FID to make the average equal to zero (Fig. 3.14). This process is

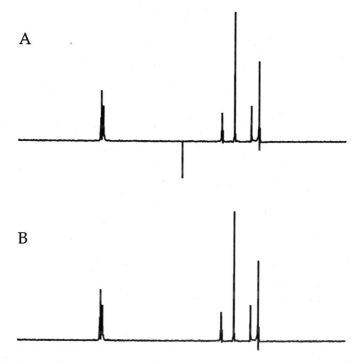

Fig. 3.14. A zero-frequency spike in the DFT spectrum due to a constant offset in the FID (**A**) can be removed by subtracting the offset (**B**).

sometimes referred to as removing a DC offset. A common practice is to subtract the average of the last eighth of the data—the real and imaginary averages are computed separately—from the entire FID. It is a good idea to remove the DC offset before applying apodization. As with quad images, the practice of alternating the signal routing and cycling the phase usually suffices to eliminate a DC offset.

3.15.2. Baseline Curvature

If the ultimate goal is to locate the peaks in a spectrum, one also needs to be able to identify areas that are not peaks. The parts of the spectrum that are not peaks (or artifacts) form the *baseline*. In other kinds of spectroscopy the "not peak" area is referred to as the *background*. Distinguishing "peak" from "not peak" is easiest when the baseline is utterly featureless: a straight line, or flat baseline. Curved or rolling baselines not only complicate the identification of peaks, they also can present a significant source of error in quantification, as can flat baselines that are offset from zero (see Chapter 7).

As we saw in Chapter 2, the first few points of the FID are responsible for slowly varying undulations in the DFT spectrum. Consequently, curved baselines can be traced to errors in the early points of the FID. It is not uncommon for the noise level to be significantly greater for these points than for the rest of the FID, as a result of imperfections in the circuitry of the spectrometer. We'll see in Chapter 4 how to correct baseline curvature by manipulating just the initial points. An alternative is to construct a model of the baseline in the frequency domain, and subtract the model from the spectrum. This method has the advantage that the new baseline will not have a constant offset.

A particularly simple, if brutish, way of modeling the baseline is to select a well-distributed set of points that fall on the baseline and then interpolate between those points to complete the model. It may seem quaint in this age of personal digital assistants, but we find it convenient to identify the baseline points manually. The value of the baseline at each point can be estimated as the value of the spectrum at that point, or if the data are very noisy, as the average of the spectrum over some small neighborhood centered on that point. Note that the baseline level is a complex number, since the spectrum is complex.

The interpolating function for estimating the baseline between the selected points can be as simple as a straight line, or as complicated as you care to make it. A low-order polynomial has the virtue of smoothing the corners, but when noise is present, the difference between linear and polynomial interpolation is often indistinguishable. In our experience, linear interpolation is usually adequate (an example is shown in Fig. 3.15). Regardless of the interpolating function, the result is an estimate of the baseline for each frequency in the spectrum, which can be subtracted to give a nearly flat baseline.

3.15.3. Solvent Signals

A recurring problem in high resolution NMR is that of determining the spectrum of a solute present at low concentration when the solvent gives rise to peaks near those of the solute. The most important example is the study of biological macromolecules in water. Biomolecule concentrations are often limited (by solubility) to the millimolar range, while the effective concentration of water protons is around 100 molar—a factor of 10^5 more abundant. The best solutions to this dynamic range problem lie in careful experiment design. Yet even a very successful experimental method for reducing the strong solvent resonance can leave behind a residual signal strong enough to obscure useful information. Data processing can help—somewhat.

As with baseline curvature, a practical method for eliminating a strong solvent resonance is to construct a model of the offending resonance, then subtract the model from the data. The following method is due to Marion, Ikura, and Bax [10]. The basic idea is to apply a filter which suppresses frequencies away from the solvent resonance, and use the output from the filter to construct the model. The estimate of the solvent signal **s** is found by convolving the FID with a kernel **a** of length $2J + 1$:

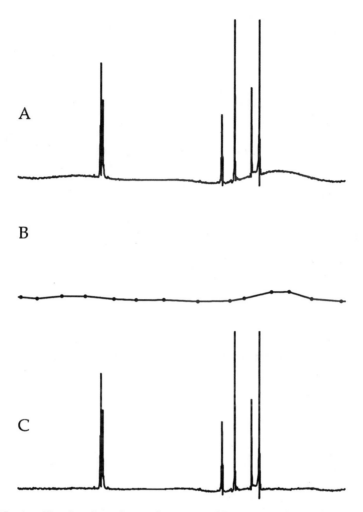

Fig. 3.15. A rolling baseline (**A**) can be corrected by constructing a piecewise linear model of the baseline (**B**), and then subtracting it point by point from the spectrum (**C**). While the result is not perfect, it is a big improvement.

$$s_k = \sum_{j=-J}^{J} a_j d_{k+j}/A, \quad \text{where} \quad A = \sum_{j=-J}^{J} a_j \qquad (3.25)$$

The width of the filter, which should be comparable to the width of the solvent resonance, is inversely proportional to the length of **a**. When the solvent is at zero frequency (the most common situation), two useful possibilities for **a** are a cosine bell,

$$a_j = \cos[j\pi/(2J + 2)] \qquad (3.26)$$

or a Gaussian,

$$a_j = e^{-4j^2/J^2} \tag{3.27}$$

Full circular convolution of **a** with the FID in Eq. (3.25) means that for $k < J$ or $k \geq M - J$, s_k would be an average of points at the end of the FID and points at the beginning. But this is not what we want. Although the DFT implicitly assumes the data are periodic, the solvent signal is not. The model function **s** can be improved by treating the beginning and the end of the FID differently; the value of s_k for $k < J$ is determined by linear extrapolation based on the values of s_k, s_{k+1}, $\ldots$, s_{k+L} (where L can be chosen empirically), and similarly for $k \geq M - J$. (In principle one could use linear prediction, discussed in Chapter 4; however in our experience LP doesn't improve the solvent model very much. Backward LP is useful when applied, as a final step, to the FID after subtracting the solvent model.) As a result of this special treatment, **s** is

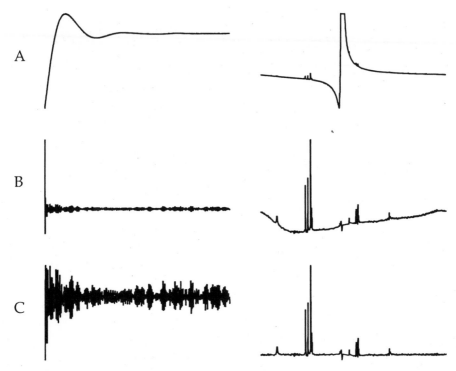

Fig. 3.16. A strong solvent signal close to zero frequency, as shown in the FID and spectrum in (**A**), can be removed by subtracting a model for the solvent signal. (**B**) shows the corrected FID and its DFT spectrum. The signal peaks close to the solvent resonance are much more visible than in (**A**). Backwards linear prediction can be used to remove the residual baseline curvature (**C**). Note that the plots are not drawn to the same scale.

not strictly speaking a filtered version of **d**, and the DFT spectrum of **s** can include the wide wings of the solvent resonance.

The parameters chosen by the user are the form of the kernel, the half-width J, and the number of points L used as the basis for the extrapolation. Once constructed, the model **s** is subtracted from **d**. Figure 3.16 shows that this procedure makes it much easier to observe peaks close to the solvent resonance. It is simple to generalize the procedure to handle strong solvent resonances that are not at zero frequency, either by first demodulating the data by the frequency of the solvent resonance and then shifting the frequencies back after subtracting the solvent model, or by using a complex-valued kernel.

3.16. DATA PROCESSING IN MULTIPLE DIMENSIONS AND HYPERCOMPLEX DATA

Lots of wonderful capabilities have resulted from the development of modern FT-NMR, but none have been quite so significant as the ability to perform multidimensional experiments. It is no exaggeration to say that without multidimensional NMR experiments, none of the applications that have fueled the explosive growth of NMR over the past two decades—including biomolecular structure determination and magnetic resonance imaging—would exist today.

For all of its importance, processing of multidimensional data requires surprisingly few concepts beyond those involved in processing one-dimensional data. An n-dimensional data set,

$$\mathbf{d}(t_1, t_2, t_3, \ldots, t_n) \quad \text{with elements} \quad d_{k_1, \ldots, k_n} \qquad (3.28)$$

consists of discrete samples of the response for each combination of the times

$$t_i = k_i \Delta t_i, \qquad k_i = 0, \ldots, M_i - 1 \qquad (3.29)$$

where the number of discrete times sampled for the time variable t_i is M_i and Δt_i is the time increment between samples for the ith dimension. The time variables represent the durations of various delays (called *evolution times*) in a pulse sequence. Sometimes the ith evolution time starts with a value $\delta_i > 0$; then Eq. (3.29) must be generalized to

$$t_i = \delta_i + k_i \Delta t_i, \qquad k_i = 0, \ldots, M_i - 1 \qquad (3.30)$$

The last time variable t_n by convention is reserved for the time between the last pulse of the sequence and collection of the data value $d(t_1, t_2, t_3, \ldots, t_n)$. The corresponding dimension is called the *acquisition dimension*, because for fixed $t_1, t_2, \ldots, t_{n-1}$, the data for all the values of t_n can be collected simply by sampling the signal following the last pulse. By contrast, the first $n - 1$ dimensions are called the *indirect dimensions*, because the only way to sample different values of those time variables is to apply pulse sequences with different delays.

The magnetization, which is the quantity measured in NMR experiments, can be viewed as a vector that rotates during each time interval. As a result, the data have both a real and an imaginary component for *each* dimension. This is a rather unusual data format; such data sets are called *hypercomplex*. A hypercomplex two-dimensional data set can be decomposed into four quadrants, each one a real, two-dimensional array (Fig. 3.17A): real-real, real-imaginary, imaginary-real, and imaginary-imaginary. A hypercomplex three-dimensional data set can likewise be decomposed into eight octants, each one a real three-dimensional array.

A
$$
t_1 \text{ real,} \quad
\begin{bmatrix}
d^{rr}_{0,0} & \cdots & d^{rr}_{0,M_2-1} \\
\vdots & & \vdots \\
d^{rr}_{M_1-1,0} & \cdots & d^{rr}_{M_1-1,M_2-1}
\end{bmatrix}
\quad
t_1 \text{ real,} \quad
\begin{bmatrix}
d^{ri}_{0,0} & \cdots & d^{ri}_{0,M_2-1} \\
\vdots & & \vdots \\
d^{ri}_{M_1-1,0} & \cdots & d^{ri}_{M_1-1,M_2-1}
\end{bmatrix}
$$
t_2 real, t_2 imaginary

$$
t_1 \text{ imaginary,} \quad
\begin{bmatrix}
d^{ir}_{0,0} & \cdots & d^{ir}_{0,M_2-1} \\
\vdots & & \vdots \\
d^{ir}_{M_1-1,0} & \cdots & d^{ir}_{M_1-1,M_2-1}
\end{bmatrix}
\quad
t_1 \text{ imaginary,} \quad
\begin{bmatrix}
d^{ii}_{0,0} & \cdots & d^{ii}_{0,M_2-1} \\
\vdots & & \vdots \\
d^{ii}_{M_1-1,0} & \cdots & d^{ii}_{M_1-1,M_2-1}
\end{bmatrix}
$$
t_2 real, t_2 imaginary

B
$$
t_1 = 0, \text{ real:} \quad \left[\left(d^{rr}_{0,0}, d^{ri}_{0,0} \right) \cdots \left(d^{rr}_{0,M_2-1}, d^{ri}_{0,M_2-1} \right) \right]
$$
$$
t_1 = 0, \text{ imaginary:} \quad \left[\left(d^{ir}_{0,0}, d^{ii}_{0,0} \right) \cdots \left(d^{ir}_{0,M_2-1}, d^{ii}_{0,M_2-1} \right) \right]
$$
$$
\vdots \qquad \qquad \vdots
$$
$$
t_1 = M_1 - 1, \text{ real:} \quad \left[\left(d^{rr}_{M_1-1,0}, d^{ri}_{M_1-1,0} \right) \cdots \left(d^{rr}_{M_1-1,M_2-1}, d^{ri}_{M_1-1,M_2-1} \right) \right]
$$
$$
t_1 = M_1 - 1, \text{ imaginary:} \quad \left[\left(d^{ir}_{M_1-1,0}, d^{ii}_{M_1-1,0} \right) \cdots \left(d^{ir}_{M_1-1,M_2-1}, d^{ii}_{M_1-1,M_2-1} \right) \right]
$$

C
$$
w_2 = 0, \text{ real} \quad w_2 = 0, \text{ imaginary} \cdots \quad w_2 = N_2 - 1, \text{ real} \quad w_2 = N_2 - 1, \text{ imaginary}
$$
$$
\begin{bmatrix} \left(d^{rr}_{0,0}, d^{ir}_{0,0} \right) \\ \vdots \\ \left(d^{rr}_{M_1-1,0}, d^{ir}_{M_1-1,0} \right) \end{bmatrix}
\begin{bmatrix} \left(d^{ri}_{0,0}, d^{ii}_{0,0} \right) \\ \vdots \\ \left(d^{ri}_{M_1-1,0}, d^{ii}_{M_1-1,0} \right) \end{bmatrix}
\cdots
\begin{bmatrix} \left(d^{rr}_{0,N_2-1}, d^{ir}_{0,N_2-1} \right) \\ \vdots \\ \left(d^{rr}_{M_1-1,N_2-1}, d^{ir}_{M_1-1,N_2-1} \right) \end{bmatrix}
\begin{bmatrix} \left(d^{ri}_{0,N_2-1}, d^{ii}_{0,N_2-1} \right) \\ \vdots \\ \left(d^{ri}_{M_1-1,N_2-1}, d^{ii}_{M_1-1,N_2-1} \right) \end{bmatrix}
$$

Fig. 3.17. A two-dimensional hypercomplex data set contains four quadrants, each an $(M_1 \times M_2)$ matrix (**A**). Unlike one-dimensional data, for which each point has two components (real and imaginary), here each point d_{mn} has four components: d^{rr}_{mn}, d^{ri}_{mn}, d^{ir}_{mn}, and d^{ii}_{mn}. To perform processing along the t_2 dimension, the data are arranged as a series of $2M_1$ one-dimensional complex row vectors of length M_2 (**B**). Each complex element consists of the corresponding real and imaginary components in t_2. After processing, each row vector has length N_2. Processing along the t_1 dimension requires that the data be arranged as $2N_2$ complex column vectors of length M_1, composed of corresponding real and imaginary components in t_1 (**C**).

Processing a multidimensional data set $\mathbf{d}(t_1, t_2, t_3, \ldots, t_n)$ to obtain a multidimensional spectrum $\mathbf{f}(w_1, w_2, w_3, \ldots, w_n)$ is accomplished by applying a series of one-dimensional DFTs along each of the dimensions. Processing in the ith dimension entails constructing a collection of vectors of length M_i, each consisting of the data values with t_i varying and the other index variables held fixed. These vectors are processed independently as ordinary one-dimensional data sets (see Fig. 3.17B and C). Zero-filling is usually part of the processing, so the final length of the vectors will be $N_i > M_i$. Since apodization, Fourier transformation, and most of the other common data processing steps are linear, the order in which the dimensions are processed will not affect the final spectrum. (Mathematically, linear operations applied along different dimensions *commute*.) However, artifacts such as strong solvent peaks are often easier to remove in the acquisition dimension, so that dimension is usually processed first. During intermediate steps of the processing, the data will be in the time domain for some dimensions and the frequency domain for others. This kind of data set is called an *interferogram*. We will denote dimensions in the frequency domain by writing w for the index; for example, $\mathbf{d}(t_1, w_2)$ represents a data set which has been processed along the second dimension, but not the first. The ith dimension is called t_i when it is in the time domain, and w_i (or f_i) when it is in the frequency domain. A typical flow diagram for the steps in processing a two-dimensional data set is shown in Figure 3.18, and a schematic diagram of the appearance of the data at each step is shown in Figure 3.19.

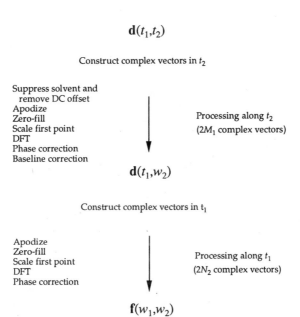

Fig. 3.18. Typical processing of a two-dimensional data set involves applying various one-dimensional operations, first along the t_2 dimension and then along the t_1 dimension. (Baseline correction is usually not needed for the indirect dimensions of a multidimensional experiment.)

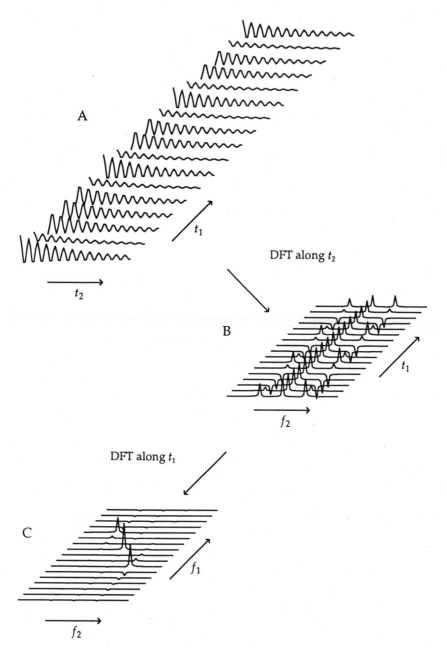

Fig. 3.19. A schematic illustration of the real-real quadrant of a two-dimensional data set is shown as it appears in the time domain (**A**), following processing along the t_2 dimension (the t_1–f_2 interferogram) (**B**), and following processing along t_1 to yield the final spectrum (**C**).

3.17. CHARACTERISTICS OF NMR SIGNALS IN MULTIPLE DIMENSIONS

We saw earlier that an FID can be described as a sum of exponentially decaying sinusoids. The same principle applies to multidimensional NMR data, but the simple formula given by Eq. (3.2) is no longer sufficient. Separate decay rates, phases, and frequencies are required for each dimension, reflecting the differing behavior of the nuclear magnetization during the different evolution periods. The projection onto the jth dimension, s_j, of a multidimensional sinusoid s can be written as:

$$s_j(t) = e^{i\phi_j} e^{-t_j/\tau_j} e^{2\pi i t_j f_j}$$

$$= e^{-t_j/\tau_j}[\cos(\phi_j + 2\pi t_j f_j) + i \sin(\phi_j + 2\pi t_j f_j)]$$

(3.31)

Each component of the full multidimensional sinusoid is the product of the corresponding components of the s_j's times a single, fixed amplitude A. Thus for a two-dimensional sinusoid, we have:

$$s^{rr}(t_1, t_2) = A \cdot \text{real}[s_1(t_1)]\text{real}[s_2(t_2)]$$

$$s^{ri}(t_1, t_2) = A \cdot \text{real}[s_1(t_1)]\text{imag}[s_2(t_2)]$$

$$s^{ir}(t_1, t_2) = A \cdot \text{imag}[s_1(t_1)]\text{real}[s_2(t_2)]$$

$$s^{ii}(t_1, t_2) = A \cdot \text{imag}[s_1(t_1)]\text{imag}[s_2(t_2)]$$

(3.32)

It is clear how to generalize this formulation to higher dimensions.

The final data set is simply a sum of sinusoids, each representing a single peak in the spectrum, plus noise. There is no deep conceptual difference between this description and the one-dimensional account using Eq. (3.2); the bookkeeping is just more tedious.

3.18. FIRST-POINT CORRECTION

We saw in Chapter 2 that one of the differences between the discrete and continuous Fourier transforms of a decaying sinusoid is a constant vertical offset. In one-dimensional spectra this rarely presents a problem. However, for multidimensional data, the amount of this offset will vary from row to row of the data matrix, leading to spurious frequency components in the indirect dimensions that appear as "ridges", called t_1-*ridges* in the case of a two-dimensional experiment (Fig. 3.20A). Since the vertical level of a spectrum is determined by the first point of the corresponding time series, these ridges can be eliminated by adjusting the first data point of each row in the time domain.

The appropriate adjustment (assuming the signal has decayed to zero by the end of the sampling period) is to multiply the first point by 0.5. This adjustment

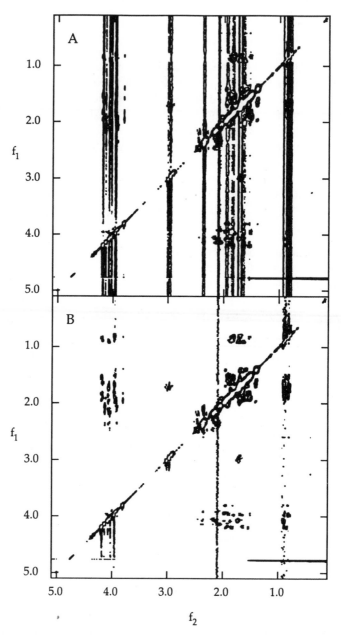

Fig. 3.20. Ridges due to baseline offset in t_1 (**A**) can be removed by multiplication of the first point of each row in t_2 by 0.5 prior to Fourier transformation (**B**).

can be rationalized in a couple of ways: One is to remember the periodicity of the DFT. A large value for the first data point represents a significant discontinuity on going from a small value at the very end of the data. Replacing the first point by the average of the first and last points will minimize the discontinuity; the average is close to one half the initial data value. A better explanation can be found in the discussion of the difference between the continuous and discrete Fourier transforms in Chapter 2 [following Eq. (2.44)]: Multiplying the first point by 0.5 precisely cancels the baseline offset inherent in the DFT of an exponential decay. Figure 3.20B shows the effect of scaling the first point (in t_2) prior to Fourier transformation.

An alternative to scaling the first point is to apply baseline correction to each vector in the acquisition dimension before processing the indirect dimensions. In addition to flattening the baseline, this adjusts the baseline level to zero for each vector, removing any offset that might be modulated in the indirect dimensions.

3.19. QUADRATURE DETECTION IN INDIRECT DIMENSIONS

Simultaneous detection of two different components of the signal isn't possible for the $n - 1$ indirect dimensions of an n-dimensional experiment. They can be detected separately, however. The *States-Haberkorn-Ruben* [11] method of data collection (also called the hypercomplex method) employs this technique, using 2^{n-1} pulse sequences to obtain the real and imaginary components in the indirect dimensions. For example, in two-dimensional experiments, two pulse sequences are required to collect all the data points for each t_1 value: one for the real component and one for the imaginary component in t_1. Each pulse sequence yields exactly one of the row vectors shown in Figure 3.17B.

An alternative is to use TPPI in the indirect dimensions, which is sometimes easier to implement. When TPPI is used in an indirect dimension, the data have only a real component in that dimension, since TPPI is a single-channel technique. Of course, once the data have been Fourier transformed, they become complex. Regardless of the choice of data collection scheme for the indirect dimensions, the final spectrum is hypercomplex, but only the real component of the spectrum is presented.

There is another quadrature detection scheme called *States-TPPI*. It is a sort of hybrid method, but the data it yields can be processed in exactly the same way as data collected with the States-Haberkorn-Ruben method.

3.20. PHASE CORRECTION AND DELIBERATE UNDER-SAMPLING IN INDIRECT DIMENSIONS

Each dimension of a multidimensional spectrum is phase corrected in the same way as a one-dimensional spectrum. The correction factors are usually the same for all the vectors along a particular dimension, so there is a single constant and linear phase correction for each dimension. As a special bonus, there usually is no need for a constant correction in the indirect dimensions. The linear correction in the ith indirect dimension can be calculated from the evolution time δ_i of the first point:

$$\text{(linear phase correction)}_i = -360° \; (\delta_i/\Delta t_i) \qquad (3.33)$$

A useful trick for improving digital resolution in indirect dimensions is to deliberately increase Δt_i, called *undersampling*. For fixed N_i, the resolution $1/N_i\Delta t_i$ is better, but the spectral width $1/\Delta t_i$ is reduced, which can cause aliasing. This isn't a problem if there is some way to distinguish which peaks have been aliased. One possibility is to choose the spectral width and the carrier frequency so that aliased peaks only appear in an otherwise blank region of the spectrum. Another technique is to use an initial evolution time $\delta_i = \Delta t_i/2$, which results in a 180° linear phase shift in the spectrum. The peaks will receive a phase shift corresponding to their true nonaliased frequencies; peaks that are aliased once will therefore undergo a phase shift 180° larger than nonaliased peaks. Following the linear phase correction, the aliased peaks will have opposite sign from nonaliased peaks, making them easy to distinguish. Peaks that are twice aliased will have an additional phase shift of 360° relative to the nonaliased peaks, so they will appear with the same sign.

3.21. SPECIAL PROCESSING REQUIREMENTS: P.COSY AND ENHANCED SENSITIVITY

There are a few multidimensional experiments that have especially unusual data processing requirements. Some examples are the two-dimensional P.COSY (purged correlation spectroscopy) experiment and the Cavanagh-Rance sensitivity-enhancement procedure (sometimes called PEP, for preservation of equivalent pathways). Although we won't go into the theory behind these techniques, we will describe how to handle the data they generate.

The P.COSY experiment is like a regular COSY, except that a special reference array is subtracted from the main data array at the start of processing [12]. The effect is to change the peak shape of the broad diagonal peaks of the COSY spectrum from dispersive to absorptive form, which reduces the extent of overlap with cross peaks. The reference array **d** is special because each row is simply a shifted version of the first. That is,

$$d(k_1\Delta t_1, k_2\Delta t_2) = d(0, k_1\Delta t_1 + k_2\Delta t_2) \qquad (3.34)$$

If the same spectral width is used in w_1 and w_2, so that $\Delta t_1 = \Delta t_2$, the entire reference data **d'** can be collected in a single one-dimensional experiment.

For the real part in t_1: $\qquad\qquad d_{k_1,k_2} = d'_{k_1+k_2}$

For the imaginary part in t_1: $\quad d_{k_1,k_2} = \pm\, id'_{k_1+k_2}$

$$(3.35)$$

The number of points in $\mathbf{d}'$ must be at least as large as $M_1 + M_2$, and whether the imaginary part requires a multiplicaiton by $+i$ or $-i$ depends on the details of the experiment.

In practice, the $\mathbf{d}'$ data set generally includes a large amount of signal summation to improve its S/N; that is, the data set is collected repeatedly, N_s times, and the results are added together. The S/N increases as the square root of N_s, and since acquiring one-dimensional data is so quick compared to two-dimensional data, the cost of this repeated acquisition is minimal. Therefore, to process the P.COSY experiment all you need to do is subtract $\mathbf{d}$ (element by element) from the main data array, where $\mathbf{d}$ is computed from $\mathbf{d}'$ using Eq. (3.35) and division by N_s (to scale the values back to the proper level). We need to add a disclaimer: If TPPI is used for sign discrimination in w_1 rather than States-Haberkorn-Ruben, the procedure becomes rather more complicated.

The Cavanagh-Rance sensitivity-enhancement procedure exploits the fact that in certain NMR experiments, important magnetization components are present along two axes simultaneously [13]. By properly partitioning the acquired data into two parts, it is possible to capture the signal components along both axes, although one component will be out of phase with respect to the other. If the noise in the two parts is uncorrelated, adding them together will then yield a spectrum with an improvement of $\sqrt{2}$ in S/N. For simplicity we will just describe the processing appropriate for a two-dimensional TOCSY (total correlation spectroscopy) experiment, but other experiments require very similar treatment.

Because of the partition mentioned above, there are two data sets to be processed: $\mathbf{d}^A$ and $\mathbf{d}^B$. The first step is to compute the sum and the difference of these two sets, element by element, yielding $\mathbf{d}^+$ and $\mathbf{d}^-$ respectively. The difference $\mathbf{d}^-$ contains the out-of-phase magnetization component, so it must be phased by $90°$ in both t_1 and t_2. The final data set $\mathbf{d}$ is then the sum of $\mathbf{d}^+$ and the properly phased $\mathbf{d}^-$ (or possibly the difference, depending on the experimental details). It is possible to perform all these manipulations in the frequency domain instead, if you choose. Just process $\mathbf{d}^A$ and $\mathbf{d}^B$ as normal TOCSY experiments to get spectra $\mathbf{f}^A$ and $\mathbf{f}^B$, then compute $\mathbf{f}^+$, $\mathbf{f}^-$, and finally $\mathbf{f}$ in the way described above.

3.22. SYMMETRIZATION

In theory, many multidimensional spectra should contain additional symmetries beyond those inherent in the DFT. The most common is reflection symmetry about the diagonal ($w_1 = w_2$) in two-dimensional spectra. A simple means for minimizing many artifacts and improving S/N in these spectra is to enforce the symmetry. For example, assuming the discrete frequencies in w_1 match those in w_2, the symmetrization can be accomplished by replacing spectrum values $f(w_1, w_2)$ with the average of $f(w_1, w_2)$ and $f(w_2, w_1)$. Alternatively, both can be replaced with the smaller of the two. Either method can improve the S/N, but each can also lead to distortion of line shapes and even propagation of artifacts. If the frequencies in w_1 do not match those in w_2, symmetrization is still possible, but it requires interpolation between the frequency values. There are many other methods for improving S/N that do not have the drawbacks of symmetrization. In general, we recommend against using it.

TO READ FURTHER

A prime source for practical details on NMR data processing is the manual for the software package you use. In some instances you may find that you like the way one package works, but like the documentation for another package better. Don't hesitate to look at all of the software documentation that is available to you, whether you prefer to use one package to the exclusion of all others or not.

References 14 through 19 are books on experimental NMR that include useful discussions of data processing. Reference 20 contains a thorough analysis of the properties of the DFT and its applications in NMR and other areas.

4

LINEAR PREDICTION

A general characteristic of spectrum analysis is that long time series are required for high frequency resolution. As a result, extrapolation of FIDs is very important in NMR data processing. The simplest form of extrapolation is zero-filling, which we have seen has limitations. A much better method is *linear prediction* (LP). In addition to providing more accurate extrapolation, it also has important applications to modeling FIDs as sums of decaying sinusoids, which is one form of parametric spectrum analysis. We should warn the reader that this chapter contains some heavy-duty mathematics. If you find yourself getting snowed by the equations, press on anyway; you should still be able to get a good sense of the strengths and limitations of LP.

4.1. DEFINITION

The theoretical underpinnings of the LP method date back to the work of Baron Prony, two centuries ago. The method is based on a single fundamental principle that holds true for many different kinds of data (including free induction decays in NMR): Each value in the time series can be represented as some fixed linear combination of the immediately preceding values. Expressed mathematically, the basic LP equation says

$$d_k = \sum_{j=1}^{m} a_j d_{k-j}, \qquad k = m, \ldots, M - 1 \tag{4.1}$$

In this formula, the values a_j are the *LP coefficients*, sometimes referred to as the *LP prediction filter*, and their number m is called the *order* of the prediction filter. Given a time series and LP coefficients that make the equation true, it is immediately clear that we can extrapolate the series beyond M points. All we need do is apply Eq. (4.1) with $k = M$ to predict a value for d_M; then this value can be used to help determine a value for d_{M+1}, and so on for as long as we desire. Even if the LP equation is only approximately true for the original time series, it can still be used to compute an approximate extrapolation— although the errors will naturally grow larger as more points are predicted.

Equation (4.1) is sometimes called the *forward* LP equation, because it gives a way to extrapolate a time series in the forward direction, predicting later values in terms of earlier ones. But the linear prediction principle can be applied just as well in the *backward* direction. The fundamental backward LP equation is

$$d_k = \sum_{j=1}^{m} b_j d_{k+j}, \qquad k = 0, \ldots, M - m - 1, \qquad (4.2)$$

which predicts each value from the immediately following ones in a time series. Equation (4.2) can be derived from Eq. (4.1) by very simple algebraic manipulation (although the values this procedure yields for the coefficients b_j generally are not the best ones to use). Both forms of LP have their advantages, and we will see examples of each kind in this chapter.

The main reasons for the success of LP are these: The class of time series that obey the LP equation coincides with the class of sums of exponentially decaying (or growing) sinusoids, and LP provides a way of *linearly* fitting a model to such a time series. The last point is especially telling, since nonlinear fitting procedures are notoriously finicky and difficult to use.

The advantages of LP extrapolation over zero-filling are fairly obvious. Assuming the prediction filter is stable (see below), the time series it generates is much more realistic than one in which the signal suddenly stops short and instantaneously decays to zero. The sinc wiggle artifacts so commonly encountered with zero-filling are greatly reduced, without the need to apply a drastic line-broadening apodization function. This property is very useful for processing multidimensional experiments, where the number of points in the indirect dimensions may be quite limited. Going in the other direction, backward LP is sometimes used to extrapolate the first few values in an FID from the remaining portion. This may seem to be a useless exercise, but when the initial points have been corrupted (for example, by ringing in the analog circuitry of a spectrometer) this procedure can fix them up and thereby help eliminate baseline curvature. Examples are shown in Figure 4.1 and Figure 3.16.

The application of LP to modeling NMR data as a sum of decaying sinusoids is more interesting. (It is also harder to perform than extrapolation.) In principle this approach obviates the need to produce a spectrum at all, or to go through the tedious work of measuring frequencies, quantifying intensities, and so forth. Instead, all the important information can be obtained directly from the parameters of the model. We will refer to this application as *parametric* LP, in

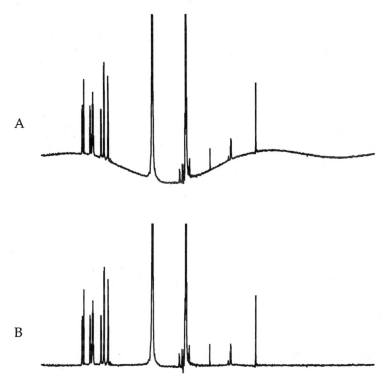

Fig. 4.1. Baseline curvature due to corruption of the first few points of the FID (**A**) can be reduced by using backwards linear prediction to recalculate the values of those points (**B**).

contrast to LP *extrapolation* as explained above. Much of the development of algorithms for performing LP calculations has been carried out by people primarily interested in parametric LP. Consequently, in the literature such names as LP-SVD and LP-TLS are nearly always used in connection with this form; however, there is no reason why these algorithms should not also be used for LP extrapolation. (The mysteries of these strange-sounding acronyms will be revealed below; for now, suffice it to say that the letters following ''LP-'' refer to the method used to perform the LP analysis.)

In addition, there are some more peripheral techniques associated with LP, such as LP-HSVD, LP-Z, and Burg MEM, that do not quite fit into our classification scheme. We will discuss these topics toward the end of the chapter.

4.2. THE THEORETICAL ANALYSIS

For what signals **d** can we find coefficients **a** that make the LP equation [Eq. (4.1)] true? Broadly speaking, any signal that is a sum of exponentially decaying (or growing) sinusoids will work. This answer is rather vague, however, particularly since we already know that *any* signal can be represented as a sum of sinusoids. Here is a more precise answer:

> Any complex-valued signal that is a sum of m exponentially decaying (or growing) sinusoids will satisfy an LP equation with a prediction order equal to m and complex coefficients. Any real-valued signal that is a sum of m exponentially decaying (or growing) sinusoids will satisfy an LP equation with a prediction order equal to $2m$ and real coefficients.

We will only consider the complex-valued case, because it is easier to explain. To start with the very simplest possibility, consider a signal that consists of just a single decaying sinusoid:

$$d_k = (Ae^{i\phi})e^{-\pi Lk\Delta t}e^{2\pi ik\Delta tf} \tag{4.3}$$

where A is the amplitude, ϕ is the phase, Δt is the sampling interval, L is the line width, and f is the frequency. This equation can be rewritten as

$$d_k = \alpha\beta^k, \quad \text{where} \quad \alpha = Ae^{i\phi} \quad \text{and} \quad \beta = e^{-\pi L\Delta t + 2\pi i\Delta tf} \tag{4.4}$$

Using this form of the equation, it is evident that **d** will satisfy the LP equation $d_k = \beta d_{k-1}$, which has order 1. The next question is: Which prediction filters of order m will also work for this same **d**? Knowing the form of **d**, we can use Eq. (4.4) to rewrite Eq. (4.1) as follows:

$$\alpha\beta^k = \sum_{j=1}^{m} a_j(\alpha\beta^{k-j}) \tag{4.5}$$

Divide through by $\alpha\beta^{k-m}$ to obtain

$$\beta^m = \sum_{j=1}^{m} a_j\beta^{m-j} \tag{4.6}$$

This will hold if, and only if, β is a root of the *characteristic polynomial $P(z)$* given by:

$$P(z) = z^m - a_1z^{m-1} - \cdots \quad _{-1}z - a_m \tag{4.7}$$

For any set of decaying sinusoids with line widths L_j, frequencies f_j, and corresponding quantities β_j (for $j = 1, \ldots, m$), there is a characteristic polynomial of order m whose roots are β_1 through β_m:

$$P(z) = (z - \beta_1)(z - \beta_2) \cdots (z - \beta_m) \tag{4.8}$$

The coefficients of this polynomial form an LP prediction filter of order m that is satisfied by each of the sinusoids and hence also by their sum, since the LP equation is linear.

This analysis actually tells us something extra. Given a set of LP coefficients, the roots of the characteristic polynomial immediately yield the possible line widths and frequencies of any signal satisfying the corresponding LP equation. From Eq. (4.4), the line width and frequency of the signal component corresponding to the root β are given by:

$$L = \frac{-1}{\pi \Delta t} \ln |\beta| \quad \text{and} \quad f = \frac{1}{2 \pi \Delta t} \text{phase}(\beta) \qquad (4.9)$$

where phase(β) is arctan[real(β)/imag(β)], measured in radians. Confusingly, the *roots* of the characteristic polynomial are sometimes referred to in the literature as the signal *poles*; this usage derives from the autoregressive (AR) model in filtering theory, where the spectral response of a linear prediction filter is given by $1/P(z)$.

For the sake of completeness, we should point out that when the characteristic polynomial has degenerate roots, the situation is more complicated than indicated above. If β is a root with multiplicity n, the signal component with frequency and decay rate corresponding to β can have the form of a decaying sinusoid, as before, but multiplied by an arbitrary polynomial of order $n - 1$. This possibility is usually ignored, on the grounds that it is quite rare for a polynomial to have degenerate roots—especially one that has been corrupted by noise.

4.3. THE STABILITY REQUIREMENT

The linear prediction method does not directly distinguish between exponentially decaying and exponentially growing signals. That is, the value of L in Eq. (4.3) could be negative, and the remainder of the analysis would proceed unchanged. An LP filter that is satisfied by exponentially growing signals is called *unstable*; when it is used for extrapolation of a time series, the computed values quickly become extremely large and result in numerical overflow. There is generally no reason to use such filters for analysis of NMR data, since almost all NMR signals decay (the major exception being signals in a constant-time dimension, which neither decay nor grow).

The moral: Whenever LP is used, it is vital to insure that the prediction filter does not admit growing signals. The line widths that an LP filter will generate are given by the roots of the characteristic polynomial. Since decaying signals have positive line widths and growing signals have negative line widths, Eq. (4.9) shows that for forward linear prediction all of the roots of the characteristic polynomial should lie within the unit circle in the complex plane: $|\beta| < 1$. For backward linear prediction exactly the opposite is true, because a decaying signal appears to be growing when viewed backward, and vice versa.

In practice, some sort of regularization procedure is needed to eliminate the unwanted roots. Perhaps the most common technique is to solve for all the roots of $P(z)$, replace each root lying outside the unit circle with its image reflected about the circle (that is, replace β with $1/\beta*$, where $\beta*$ is the complex conjugate of β), then recombine the altered roots to get a revised polynomial $P'(z)$ whose coefficients give the regularized prediction filter. Another possibility would be to replace an unwanted root with the nearest point on the unit circle (replace β with $\beta/|\beta|$). A technique that should *not* be used is simply to factor out the unwanted roots, yielding a lower-order filter. The results of doing so are uniformly awful.

These regularization procedures are not well suited for data in a constant-time dimension. Such signals have essentially no decay, and so all one can say is that the valid roots lie on or near the unit circle. Probably the best thing to do is move all the roots to the unit circle (or at least move the ones that are not already very close).

4.4. THE PARAMETRIC APPROACH

We are given a time series **d**, and we want to model it as a sum of m decaying sinusoids. In the notation of Eq. (4.4), we want to find α_j and β_j for $j = 1, \ldots, m$, satisfying the following system of equations:

$$d_k = \alpha_1 \beta_1^k + \cdots + \alpha_m \beta_m^k, \qquad k = 0, \ldots, M - 1 \qquad (4.10)$$

The LP-HSVD method provides values for the β's directly, whereas the other LP algorithms merely provide LP coefficients and leave it up to us to find the β's by factoring the characteristic polynomial. (These algorithms are described in detail below.) At any rate, suppose we have the β's, and the ones corresponding to growing signal components have already been regularized. The α's can be found by solving the system in Eq. (4.10), which is just a set of linear equations in $\alpha_1, \ldots, \alpha_m$ that can be treated as an ordinary least-squares problem. Any of the standard techniques will work here. Finally, the amplitudes and phases of the sinusoids can be recovered from the α values: The amplitude of the jth signal component is $|\alpha_j|$ and the phase is equal to the complex phase of α_j.

With the complete spectral parameters available, it is possible to calculate a synthetic spectrum, by adding together an assortment of peaks with the appropriate frequencies, amplitudes, and so on. Alternatively, one could skip entirely the step of presenting a spectrum and proceed directly with the analysis of the peaks. After all, the main use one makes of a spectrum is to measure the characteristics of the peaks; with the parametric method, all that information is already available.

You definitely need to take care when using the parametric approach. We will go into more detail later in the section on performance; for now, suffice it to say that the LP procedure doesn't always find all the peaks. If you rely solely on the list of parameters produced by this procedure, without some sort of secondary check (say a Fourier transform of the signal), there is no way to know whether all the important components of the signal have been included properly. In addition, a parametric analysis will often decompose a single peak in the spectrum into a set of closely spaced frequency components, with no indication that they should be treated as a single entity. On the other hand, when applied with due care to data with a sufficiently high S/N, the LP procedure can detect frequency components even in the face of large variations in dynamic range.

4.5. APPLYING LINEAR PREDICTION

Starting with a time series, such as an FID, the big problem is to find a set of forward or backward LP coefficients for which the time series satisfies the appropriate LP equation. Once a good set of coefficients has been found, the rest of the process—extrapolation or parametric fitting—is fairly straightforward. Unfortunately, there are several obstacles to be surmounted in the course of determining the coefficients.

1. Actual NMR signals are not perfect exponentially decaying sinusoids. Owing to instrumental imperfections such as magnetic field inhomogeneity, they may possess a sizeable Gaussian character or be distorted in other ways. In addition, complicated pulse sequences can excite higher-order magnetization terms that have a multiexponential decay pattern. (Have you ever examined the envelope of an FID from a TOCSY experiment?) In practice this is not a serious problem, because the signal decay is usually sufficiently close to exponential that an LP filter can match it fairly well. Alternatively, if a component is highly nonexponential, LP will yield a close approximation consisting of overlapping Lorentzian peaks.

2. In real experiments, the number of peaks (and hence the order of the LP prediction filter) is not known in advance. Again, this problem is not grave, because it is usually possible to estimate an upper limit to the number of peaks and use a larger filter order. The extraneous components can be detected and removed if they refer to exponentially growing signals or if they have very low intensity.

3. Real experimental data are always contaminated by noise. This is the most serious difficulty. In the presence of noise, the components detected by the prediction filter quite frequently are merely associated with the noise rather than being genuine signal peaks. The recommended approach is to use an LP filter order much higher than the expected number of peaks, depending on the S/N.

There are other, purely practical problems with solving for the LP coefficients, mainly because the procedure requires handling linear systems of equations with possibly thousands of terms (or finding the roots of polynomials of similarly high order). These problems are rather mundane, and can be dealt with by using suitable algorithms on a sufficiently powerful computer. As an example, we will present below an accurate method for finding the roots of high-order complex polynomials.

Of greater interest are the problems arising from the need to drastically increase the filter order for proper handling of noisy data. As can be seen from the LP equations [Eqs. (4.1) and (4.2)], if M is the total number of data points and m is the filter order, the total number of equations that the LP coefficients must satisfy is $M - m$. For the coefficients to be uniquely determined, m must be no larger than $M/2$. In fact, the presence of noise means that it is better for the coefficients to be overdetermined, so m is usually chosen to be no larger than $M/3$. Of course, for complicated signals such as arise in NMR experiments on polypeptides or nucleic acids, this conflicts with the requirement that m be much higher than the number of peaks, since the total number of data points can be quite limited.

Some sort of compromise is needed. As a general rule, parametric LP and very high filter orders are used in analysis of high-resolution one-dimensional experiments, where many thousands of data points can easily be collected. They are well suited, for instance, to handling data from ^{13}C or ^{31}P experiments. Conversely, in multidimensional experiments where the number of data points in the indirect dimensions may be rather small, LP is most often used just for extrapolation. With only a small filter order, it is possible that not all the components in the signal will be detected. However, a conservative extrapolation to no more than two or three times the length of the input data will not cause any major problems, especially if the extrapolated signal is then apodized, which will de-emphasize the added portion in relation to the original signal.

It is possible to use parametric LP for the complete analysis of a multidimensional data set. The technique is not in common practice, however, and we will not describe it here (for details, see reference 21). In contrast, LP extrapolation of multidimensional data is quite common, and the application is very straightforward. The simplest possibility is to independently extrapolate each t_1 vector of a two-dimensional experiment, after the t_2 vectors have undergone Fourier transformation. Although this procedure does not take into account similarities between vectors, it has the advantage that each t_1 vector will contain a relatively small set of frequency components and, therefore, will only require a low filter order.

A bewildering variety of approaches for finding LP coefficients has been published. These techniques are equally applicable to solving either the forward or backward LP equations. Both forms have their uses; for example, the forward LP equation is the appropriate form for extrapolating a time series. In parametric LP either type would yield the frequencies and decay rates, but here backward LP has an advantage. The reason comes out of work by Kumaresan [22].

Kumaresan showed that when the least-squares method (see below) is used to find the coefficients of an LP filter of order m for a time series that consists of fewer than m sinusoids and no noise, the "extraneous" roots of the characteristic polynomial (that is, the roots that do not correspond to actual signal components of the time series) will all lie inside the unit circle in the complex plane, roughly evenly dispersed around the circle. Now for backward LP, roots inside the unit circle indicate exponentially *growing*, rather than *decaying*, signals and so the extraneous roots are easily distinguished from the genuine roots, which must all lie outside the unit circle. For forward LP there is no such easy distinction, since both sets of roots will be inside the unit circle. Of course, in real applications of LP there always is noise, and methods other than least-squares may be used to find the coefficients. Nevertheless, it is often assumed that the extraneous roots will probably still tend to fall inside the unit circle. Since other requirements force the filter orders to be large, the ability to detect and eliminate quickly many of the extraneous roots means that backward LP is the preferred method. It is always possible to calculate both forward and backward prediction filters, factor the two characteristic polynomials, and then accept as valid only those roots which appear in both. This trick can be useful as an additional method for discriminating between genuine and extraneous roots. (Actually, if β is a root of the forward characteristic polynomial, the corresponding root of the backward polynomial is $1/\beta$.) Zhu and Bax recommend averaging the forward and backward values of β to reduce errors [23].

4.6. DETERMINING THE LP COEFFICIENTS

For the sake of exposition, we will concentrate on the forward LP equation. Of course, these techniques can be used just as well to solve the backward LP problem.

4.6.1. Least-Squares

Expanding the forward LP equation [Eq. (4.1)] produces

$$
\begin{aligned}
d_0 a_m + d_1 a_{m-1} + \cdots + d_{m-1} a_1 &= d_m \\
d_1 a_m + d_2 a_{m-1} + \cdots + d_m a_1 &= d_{m+1} \\
&\vdots \\
d_{L-1} a_m + d_L a_{m-1} + \cdots + d_{M-2} a_1 &= d_{M-1}
\end{aligned}
\tag{4.11}
$$

where $L = M - m$. In matrix form, this can be written as $\mathbf{D a} = \mathbf{d}'$, where

$$
\mathbf{D} = \begin{bmatrix}
d_0 & d_1 & \cdots & d_{m-1} \\
d_1 & d_2 & \cdots & d_m \\
& \vdots & & \\
d_{L-1} & d_L & \cdots & d_{M-2}
\end{bmatrix},
\tag{4.12}
$$

$$\mathbf{a} = \begin{bmatrix} a_m \\ a_{m-1} \\ \vdots \\ a_1 \end{bmatrix}, \quad \text{and} \quad \mathbf{d}' = \begin{bmatrix} d_m \\ d_{m+1} \\ \vdots \\ d_{M-1} \end{bmatrix}$$

Since $L > m$, this can be treated like any overdetermined set of linear equations and solved for $\mathbf{a}$ in the least-squares sense. The method is generically referred to as LP-LSQ, and the solution is given by

$$\mathbf{a} = (\mathbf{D}^\dagger \mathbf{D})^{-1} \mathbf{D}^\dagger \mathbf{d}' \tag{4.13}$$

where $\mathbf{D}^\dagger$ is the Hermitian transpose of $\mathbf{D}$ (that is, the complex conjugate of the transpose of $\mathbf{D}$).

At least three different techniques have been proposed for inverting the matrix $\mathbf{D}^\dagger \mathbf{D}$: Householder QR decomposition (QRD), Cholesky decomposition, and singular value decomposition (SVD). The first two yield similar results, although the Cholesky method is quicker to compute because it takes advantage of the Hermitian nature of $\mathbf{D}^\dagger \mathbf{D}$. Setting up this matrix in the first place can be done rapidly thanks to the regular structure of $\mathbf{D}$ (it is a *Hankel* matrix, meaning that the entries along each antidiagonal are all equal).

All three are standard methods for matrix inversion and we will not include detailed descriptions of how to compute them (an excellent discussion can be found in *Numerical Recipes in Fortran* [24], and comparisons of their performances are given in [21]). Briefly, QR decomposition involves applying a sequence of Householder transformations to factor $\mathbf{D}^\dagger \mathbf{D}$ into the product of a unitary matrix and an upper triangular matrix, each of which is easily inverted. Cholesky decomposition involves factoring $\mathbf{D}^\dagger \mathbf{D}$ into the product of a lower triangular matrix and its Hermitian transpose, which again leads to rapid inversion.

Singular value decomposition deserves a more extended discussion. The decomposition takes the form

$$\mathbf{D} = \mathbf{U} \mathbf{\Lambda} \mathbf{V}^\dagger \tag{4.14}$$

where $\mathbf{U}$ and $\mathbf{V}$ are $(L \times L)$ and $(m \times m)$ unitary matrices (which just means that $\mathbf{U}\mathbf{U}^\dagger$ and $\mathbf{V}\mathbf{V}^\dagger$ are identity matrices), and $\mathbf{\Lambda}$ is an $(L \times m)$ diagonal matrix of so-called *singular values* $\lambda_1, \ldots, \lambda_m$ (see Fig. 4.2). Each singular value corresponds to a component in $\mathbf{D}$, and the larger singular values will generally be associated with genuine signal components, while the smaller ones are associated primarily with noise components. The SVD method, therefore, retains only the p largest entries in $\mathbf{\Lambda}$, where p is the number of presumed signal components (also called the *reduced order*), and sets the $m - p$ smaller entries

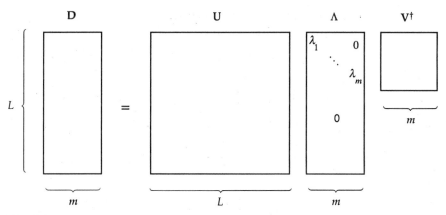

Fig. 4.2. Singular value decomposition of the $(L \times m)$ matrix **D** yields an $(L \times L)$ unitary matrix **U** multiplied by the $(L \times m)$ diagonal matrix Λ of singular values λ_1, ..., λ_m and the Hermitian transpose of the $(m \times m)$ unitary matrix **V**.

in Λ to zero before solving the least-squares problem. The solution is given by $\mathbf{a} = \mathbf{V}\Lambda^{-1}\mathbf{U}^{\dagger}\mathbf{d}'$, where Λ^{-1} is the $(m \times L)$ *pseudoinverse* of Λ: The p largest diagonal entries are replaced by their inverses and the others are replaced by zero. One way to choose p is to count how many entries are above a noise threshold. Unfortunately, the determination of where to set the noise threshold is not always clear-cut; set it too high and some real signal features can be lost, set it too low and excessive noise can corrupt the result. In lucky cases there will be an obvious gap in the range of singular values and the noise threshold can be placed in this gap; in other cases there will not. Some choice must be made, perhaps depending on the estimated S/N.

The ability to distinguish and reject noise components is definitely a strong point in favor of LP-SVD over the other least-squares methods. Several of the other algorithms discussed below incorporate an SVD step, and in each case truncating the singular values can help eliminate noise.

4.6.2. Total Least-Squares

The least-squares solution to the LP equation suffers from a subtle defect: In finding the vector **a** which minimizes the squared error $|\mathbf{Da} - \mathbf{d}'|^2$, there is an implicit assumption that only the vector $\mathbf{d}'$ includes any noise, whereas in fact **D** does as well. The total least-squares method (TLS) attempts to remedy this drawback [25]. By moving $\mathbf{d}'$ to the left-hand side of the equation $\mathbf{Da} = \mathbf{d}'$, appending it to **D** as an additional column, and appending -1 to the **a** vector, we can transform the matrix equation into

$$(\mathbf{D}|\mathbf{d}') \left(\frac{\mathbf{a}}{-1} \right) = \mathbf{0} \tag{4.15}$$

Let $\mathbf{E} = (\mathbf{D}|\mathbf{d}')$ be the augmented data matrix. In contrast to the SVD method, which makes the smallest possible adjustment to $\mathbf{d}'$ to satisfy Eq. (4.15), TLS tries to make the smallest possible change in $\mathbf{E}$. Thus $\mathbf{D}$ and $\mathbf{d}'$ are treated on a more equal footing.

In linear algebra, a matrix is viewed as describing a map from one vector space to another. In our case, $\mathbf{E}$ maps an $(m + 1)$-dimensional space to an L-dimensional space. The *rank* of $\mathbf{E}$ is the dimension of the image (i.e., the output of the map), and the *nullity* of $\mathbf{E}$ is the dimension of the null space (the vectors that $\mathbf{E}$ maps to zero). It is a theorem of linear algebra that the rank plus the nullity of a matrix is equal to the dimension of the *domain* (the input space); thus rank($\mathbf{E}$) + nullity($\mathbf{E}$) = $m + 1$. Equation (4.15) says that $\mathbf{E}$ has at least one null vector, which implies that rank($\mathbf{E}$) is less than $m + 1$. $\mathbf{E}$ is said to be *rank-deficient*, because the rank is not as large as it potentially could be. The TLS method constructs the matrix $\mathbf{E}$ and perturbs it to make it rank-deficient. The LP coefficients can then be obtained from a null vector of $\mathbf{E}$.

The best way to do this is to perform a singular value decomposition of $\mathbf{E}$:

$$\mathbf{E} = \mathbf{U}\mathbf{\Lambda}\mathbf{V}^{\dagger} \tag{4.16}$$

Assuming the singular values are in descending order ($\lambda_1 \geq \lambda_2 \geq \cdots \geq \lambda_{m+1}$), the smallest perturbation that will make $\mathbf{E}$ rank-deficient is to set the smallest one, λ_{m+1}, to zero. The corresponding singular vector is just the last column of $\mathbf{V}$, so after the appropriate normalization, the coefficient vector $\mathbf{a}$ is given by

$$\mathbf{a} = -(V_{m+1,m+1})^{-1} \begin{bmatrix} V_{1,m+1} \\ \vdots \\ V_{m,m+1} \end{bmatrix} \tag{4.17}$$

We have already noted that the reduced order p is usually smaller than the prediction order m. In the absence of noise, the augmented data matrix $\mathbf{E}$ should only have rank p. Therefore, in an attempt to better suppress the noise components of the data, the suggested approach is to set all the singular values from λ_{p+1} to λ_{m+1} to zero and then determine a set of LP coefficients from the null space of the modified matrix. The choice of which possible null vector to use is decided in the usual way: Use the one that gives rise to LP coefficients with the smallest norm. The null space is spanned by the column vectors $\mathbf{V}_{p+1}$, $\ldots$, $\mathbf{V}_{m+1}$ [that is, the $(p + 1)$ through $(m + 1)$ columns of $\mathbf{V}$]. Hence the problem is to find coefficients c_i so as to

$$\text{minimize} \quad |c_{p+1}\mathbf{V}_{p+1} + \cdots + c_{m+1}\mathbf{V}_{m+1}|$$
$$\text{subject to} \quad c_{p+1}V_{p+1,m+1} + \cdots + c_{m+1}V_{m+1,m+1} = -1 \tag{4.18}$$

The solution is to take c_i to be equal to a normalizing factor times $V^*_{i,m+1}$. The LP coefficient vector is given by

$$a_k = -\left(\sum_{j=p+1}^{m+1} |V_{j,m+1}|^2 \right)^{-1} \sum_{i=p+1}^{m+1} V^*_{i,m+1} V_{i,k} \qquad (4.19)$$

In practice, the LP-TLS method is about as quick to compute as LP-SVD, and it always gives somewhat better results. The LP-QRD and LP-Cholesky methods are quicker and do not need as much memory, and their results may be nearly as good, depending on the S/N. The loss of quality is often sufficiently slight that it is worth using the quicker methods.

4.6.3. The Hankel SVD Method

The Hankel singular value decomposition technique (HSVD), also known as the state space method, uses a rather different approach [26]. Strictly speaking, it is not a form of LP at all, since it does not produce or make use of a linear prediction filter. Instead it directly generates the set of frequencies and decay rates. As a result, it is not suited for LP extrapolation but only for parametric modeling.

LP-HSVD works by writing the data matrix **D** as a product of Vandermonde matrices and a diagonal matrix. (A *Vandermonde* matrix is one in which each row or each column consists of increasing powers of some value.) If the time series d_k really is a sum of p decaying sinusoids:

$$d_k = \alpha_1 \beta_1^k + \cdots + a_p \beta_p^k, \qquad k = 0, \ldots, M-1 \qquad (4.20)$$

then the data matrix

$$\mathbf{D} = \begin{bmatrix} d_0 & \cdots & d_{m-1} \\ & \vdots & \\ d_{L-1} & \cdots & d_{M-1} \end{bmatrix} \qquad (4.21)$$

can be decomposed in the following way:

$$\mathbf{D} = \begin{bmatrix} 1 & \cdots & 1 \\ \beta_1 & \cdots & \beta_p \\ & \vdots & \\ \beta_1^{L-1} & \cdots & \beta_p^{L-1} \end{bmatrix} \begin{bmatrix} c_1 & & 0 \\ & \ddots & \\ 0 & & c_p \end{bmatrix} \begin{bmatrix} 1 & \beta_1 & \cdots & \beta_1^{m-1} \\ & \vdots & \\ 1 & \beta_p & \cdots & \beta_p^{m-1} \end{bmatrix} \qquad (4.22)$$

where $c_i = \alpha_i\beta_i$. (This formula can be checked by direct matrix multiplication.) We will write Eq. (4.22) as $\mathbf{D} = \mathbf{X}_L\mathbf{C}\mathbf{X}_m^T$. The important fact about the matrix $\mathbf{X}_L$ is that it is a Vandermonde matrix, since the ith column consists of successive powers of β_i. To put it another way, each row is equal to the previous row multiplied by the diagonal matrix $\mathbf{Z} = diag(\beta_1, \ldots, \beta_p)$. This fact can be expressed in matrix form:

$$\mathbf{X}_L^{\text{bot}} = \mathbf{X}_L^{\text{top}}\mathbf{Z} \tag{4.23}$$

where $\mathbf{X}_L^{\text{bot}}$ is $\mathbf{X}_L$ with the top row removed (i.e., just the bottom portion of the matrix) and $\mathbf{X}_L^{\text{top}}$ is $\mathbf{X}_L$ with the bottom row removed.

The problem is to find a way of decomposing the actual data matrix into this form. Equation (4.22) rather resembles a singular value decomposition, so SVD is the first step. We obtain

$$\mathbf{D} = \mathbf{U}\mathbf{\Lambda}\mathbf{V}^{\dagger} \tag{4.24}$$

where $\mathbf{U}$ is an $(L \times L)$ unitary matrix, $\mathbf{\Lambda}$ is an $(L \times m)$ diagonal matrix, and $\mathbf{V}$ is an $(m \times m)$ unitary matrix. Just as in LP-SVD, we can assume that the p largest entries in $\mathbf{\Lambda}$ correspond to genuine signal components and that the rest only represent noise. Therefore we set all but the first p entries in $\mathbf{\Lambda}$ to zero (assuming that the diagonal elements go in order of decreasing magnitude). With all these zeros in $\mathbf{\Lambda}$, all but the first p columns of $\mathbf{U}$ and $\mathbf{V}$ are now unnecessary. So the reduced-order SVD equation becomes

$$\mathbf{D} = \mathbf{U}'\mathbf{\Lambda}'\mathbf{V}'^{\dagger} \tag{4.25}$$

where $\mathbf{U}'$ is an $(L \times p)$ submatrix of $\mathbf{U}$, $\mathbf{\Lambda}'$ is a $(p \times p)$ diagonal matrix, and $\mathbf{V}'$ is an $(m \times p)$ submatrix of $\mathbf{V}$ (see Fig. 4.3).

Putting together Eqs. (4.22) and (4.25), we obtain

$$\mathbf{D} = \mathbf{X}_L(\mathbf{C}\mathbf{X}_m^T) = \mathbf{U}'(\mathbf{\Lambda}'\mathbf{V}'^{\dagger}) \tag{4.26}$$

Here $\mathbf{D}$ is expressed in two ways as the product of an $(L \times p)$ matrix and a $(p \times m)$ matrix (the terms in parentheses). Now assuming there are p sinusoids present in the signal, the data matrix $\mathbf{D}$ has rank p. It is a simple exercise in linear algebra to show that under these conditions $\mathbf{X}_L$ and $\mathbf{U}'$ must be equivalent—that is, there is an invertible $(p \times p)$ matrix $\mathbf{Q}$ such that

$$\mathbf{X}_L = \mathbf{U}'\mathbf{Q}, \quad \text{hence} \quad \mathbf{X}_L^{\text{bot}} = \mathbf{U}'^{\text{bot}}\mathbf{Q} \quad \text{and} \quad \mathbf{X}_L^{\text{top}} = \mathbf{U}'^{\text{top}}\mathbf{Q} \tag{4.27}$$

Combining Eq. (4.27) with Eq. (4.23), we see that

$$\mathbf{U}'^{\text{bot}} = \mathbf{U}'^{\text{top}}\mathbf{Z}' \tag{4.28}$$

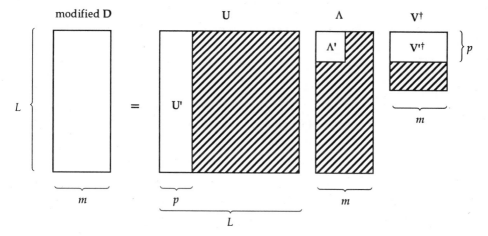

Fig. 4.3. The singular value decomposition of the data matrix $\mathbf{D}$ can be reduced to rank p by setting to zero the elements of the diagonal matrix $\boldsymbol{\Lambda}$ that lie in the shaded portion. Then the shaded portions of $\mathbf{U}$, $\boldsymbol{\Lambda}$, and $\mathbf{V}^{\dagger}$ can be eliminated, yielding the ($L \times p$) submatrix $\mathbf{U}'$, the ($p \times p$) diagonal submatrix $\boldsymbol{\Lambda}'$, and the ($p \times m$) submatrix $\mathbf{V}'^{\dagger}$.

where $\mathbf{Z}' = \mathbf{QZQ}^{-1}$. To find the β values, all that is needed is to solve Eq. (4.28) for $\mathbf{Z}'$ and then determine its eigenvalues. Since $\mathbf{Z}'$ is similar to $\mathbf{Z}$, the eigenvalues of $\mathbf{Z}'$ are the same as the entries in the diagonal matrix $\mathbf{Z}$.

Of course, Eq. (4.28) is overdetermined, and in the presence of noise it should be solved in the least-squares sense:

$$\mathbf{Z}' = (\mathbf{U}'^{\text{top}\dagger}\mathbf{U}'^{\text{top}})^{-1}\mathbf{U}'^{\text{top}\dagger}\mathbf{U}'^{\text{bot}} \tag{4.29}$$

Using the fact that the columns of $\mathbf{U}'$ are orthogonal and normalized, it is easy to check that the (i, j)-element of $\mathbf{U}'^{\text{top}\dagger}\mathbf{U}'^{\text{top}}$ is equal to $-U'^{*}_{L,i}U'_{L,j}$ if $i \neq j$, or to $1 - U'^{*}_{L,i}U'_{L,j}$ if $i = j$. In other words,

$$\mathbf{U}'^{\text{top}\dagger}\mathbf{U}'^{\text{top}} = \mathbf{I} - \mathbf{u}\mathbf{u}^{\dagger} \tag{4.30}$$

where $\mathbf{I}$ is the identity matrix and $\mathbf{u}^{\dagger}$ is the bottom row of $\mathbf{U}'$. The inverse of this matrix is given by the Sherman-Morrison formula, which yields the following simple expression for $\mathbf{Z}'$:

$$\mathbf{Z}' = \left(\mathbf{I} + \frac{\mathbf{u}\mathbf{u}^{\dagger}}{1 - \mathbf{u}^{\dagger}\mathbf{u}}\right)\mathbf{U}'^{\text{top}\dagger}\mathbf{U}'^{\text{bot}} \tag{4.31}$$

Alternatively, recognizing that both $\mathbf{U}'^{\text{bot}}$ and $\mathbf{U}'^{\text{top}}$ are subject to noise, Eq. (4.28) could be solved in the total least-squares manner.

(Exactly why this technique is called *Hankel* singular value decomposition is not clear to us. The initial data matrix does have a Hankel structure, but this is true for each of the LP algorithms we have discussed. Furthermore, the first critical step—truncating the singular values to help reduce the noise level—destroys the Hankel property.)

The advantage LP-HSVD enjoys over LP-SVD is that it does not require the factorization of a polynomial of order m to find the frequencies and decay rates, but rather the determination of the eigenvalues of a matrix of order p. If p is much smaller than m, this can be a significant savings.

4.6.4. The Minimum Variance Estimate

One of the most important steps in the SVD methods described above is the reduction of the data matrix to rank p. The justification for this is that if the time series really did contain only p components then the data matrix would have this rank, and so anything else merely represents noise. The procedures set all but the p largest singular values to zero, which effectively transforms the data matrix into the closest possible matrix of rank p.

As we just noted, this action does not maintain the Hankel structure. The resulting matrix, although it does have rank p, could never be a data matrix derived from any time series, because the entries along the antidiagonals are not constant. What we would really like to do is to make the smallest possible change to the *time series* so that the corresponding data matrix would have rank p. There is no direct procedure for doing so, but an iterative approach will work. The idea behind Cadzow's technique [27] is to compute the modified data matrix and then to average the entries along each antidiagonal. Repeated application will cause the data matrix to converge to a Hankel matrix of rank p.

It has been recommended that you use only one or two iterations of this regularization procedure, and also adjust the retained singular values so as to produce a *minimum variance* estimate of the true signal [28]. The new singular values are given by the formula,

$$\lambda_i' = \lambda_i - \sigma^2/\lambda_i \qquad (4.32)$$

where σ^2 is an estimate of the variance of the noise singular values. A simple choice for σ^2 is

$$\sigma^2 = \left(\frac{1}{m-p}\right) \sum_{j=p+1}^{m} \lambda_j^2 \qquad (4.33)$$

(Of course σ^2 must always be less than λ_p^2; otherwise the new value of λ_i' would be negative.) We will not give the full explanation of why this should yield a minimum variance estimate—or even just what that really means. Instead we simply point out that the noise, being random, is expected to contribute to *all* of the singular values, and so even the ones we retain ought to be compensated in some way to remove the extraneous contribution. The reason for the suggestion of using only a few iterations of regularization is that even these produce a noticeable improvement in the results (especially when the S/N is low), whereas repeated iterations degrade the accuracy of the final parameters. This is because the iterated Cadzow procedure does not converge to the correct value. It will yield a Hankel matrix of rank p, but not necessarily the one closest to the original noisy input.

In reference 28, it is also recommended that during the Cadzow regularization procedure, the width of the data matrix be set not to m (the rather large prediction order) but to m', a much smaller value between $p + 5$ and $2p$. This will cause the data matrix to be much more highly rectangular, and as a result the singular value decomposition can be computed more quickly. The regularized time series is then rearranged to form the ($L \times m$) data matrix for subsequent processing.

4.6.5. The Burg Method

John Burg deserves credit for being the originator of "modern" spectrum analysis. His so-called "maximum entropy method" (MEM, not to be confused with MaxEnt described in Chapter 5) was developed for use in geophysics and it quickly spread to many other fields, including NMR [29]. His technique involves finding linear prediction filters for time series data (although this was not its original purpose). Furthermore, he found a highly efficient algorithm for performing the computations needed to implement the method.

In spite of these advantages, we do not recommend using Burg's method for analysis of NMR data. There are a couple of reasons why not. First, the method was intended for use with *stationary* data, not signals that decay with time. As a result, it is biased towards smaller line widths than are actually present in the data. Second, the prediction filters that the method generates are not optimal, in the sense that they do not minimize the forward or the backward prediction errors. Consequently, parameters extracted from the characteristic polynomial are less accurate in all respects than values derived using one of the other methods listed above.

To illustrate the bias inherent in the Burg method, Figure 4.4 shows results obtained using the Burg method to extrapolate a synthetic FID consisting of four decaying sinusoids without noise. Figure 4.4A shows the original FID. In Figure 4.4B, C, and D are shown the results of extrapolating the first part of the FID using various prediction orders. None of them look very much like the true FID. In contrast, extrapolation using LP-SVD results in an FID that is virtually indistinguishable from the original.

Without going into the details of the Burg method (since we believe it should not be used), let us simply mention that some of the assumptions underlying the method are not appropriate for the kind of data encountered in NMR.

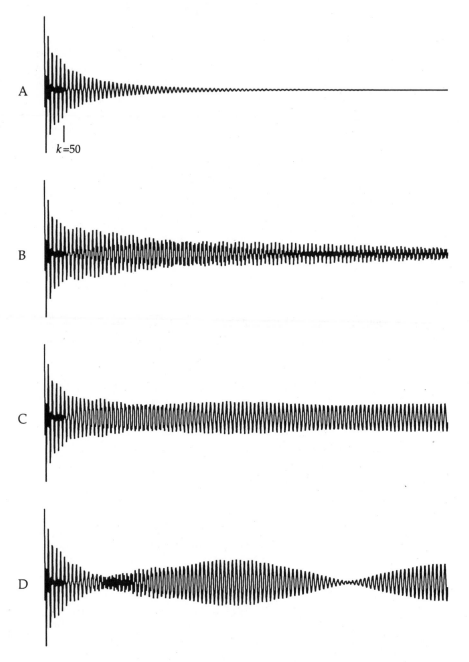

Fig. 4.4. A synthetic FID consisting of four decaying sinusoids without noise (**A**) is used as a test case for LP extrapolation via the Burg method. (**B-D**) show extrapolation from 50 to 1024 points using 4, 8, and 20 LP coefficients, respectively. Extrapolation via LP coefficients determined using LP-SVD results in a signal barely distinguishable from (**A**) (not shown).

Beware the LP routine MEMCOF found in *Numerical Recipes in Fortran* [24], which uses the Burg algorithm.

4.7. LP AND THE Z-TRANSFORM

LP-Z [30] is based on the z-transform and does not fall into the same category as any of the other methods discussed above. Instead, it generates directly the high-resolution spectrum corresponding to a time series that has been extrapolated by an LP filter. Its advantages over LP extrapolation are that it does not require the computation of any additional time series values, and it can provide the spectral intensity at any frequency, not just at multiples of $1/N\Delta t$.

The z-transform is a very close relative of the Fourier transform. Given a time series d_k ($k = 0, \ldots, \infty$), the Fourier transform f and the z-transform g are defined as:

$$f(w) = \sum_{k=0}^{\infty} d_k e^{-2\pi i k w} \quad \text{and} \quad g(z) = \sum_{k=0}^{\infty} d_k z^{-k} \tag{4.34}$$

Clearly, $f(w) = g(z)$, provided that $z = \exp(2\pi i w)$. Now suppose we have backward (instead of forward) LP coefficients b_j ($j = 1, \ldots, m$) and that **d** satisfies the backward LP equation:

$$d_k = \sum_{j=1}^{m} b_j d_{k+j}, \qquad k = 0, \ldots, \infty \tag{4.35}$$

Then we can calculate:

$$\sum_{k=0}^{\infty} d_k z^{-k} = \sum_{k=0}^{\infty} z^{-k} \sum_{j=1}^{m} b_j d_{k+j}$$

$$(\text{let } k' = k + j) \quad = \sum_{k'=1}^{m-1} d_{k'} \sum_{j=1}^{k'} b_j z^{j-k'} + \sum_{k'=m}^{\infty} d_{k'} \sum_{j=1}^{m} b_j z^{j-k'} \tag{4.36}$$

$$(\text{let } k = k') \quad = \sum_{k=0}^{m-1} d_k z^{-k} G_k(z) + \sum_{k=m}^{\infty} d_k z^{-k} G_m(z)$$

where

$$G_n(z) = \sum_{j=1}^{n} b_j z^j \quad \text{and} \quad G_0(z) = 0 \tag{4.37}$$

Rearranging Eq. (4.36) yields:

$$\sum_{k=m}^{\infty} d_k z^{-k}[1 - G_m(z)] = \sum_{k=0}^{m-1} d_k z^{-k}[G_k(z) - 1] \qquad (4.38)$$

$$\sum_{k=m}^{\infty} d_k z^{-k} = \sum_{k=0}^{m-1} d_k z^{-k}[G_k(z) - 1]/[1 - G_m(z)] \qquad (4.39)$$

$$g(z) = \sum_{k=0}^{\infty} d_k z^{-k} = \sum_{k=0}^{m-1} d_k z^{-k}[G_k(z) - G_m(z)]/[1 - G_m(z)] \qquad (4.40)$$

Arbitrary values of z can be substituted into Eq. (4.40), thereby allowing the spectrum to be computed for any frequency. By the way, notice that the expression $[1 - G_m(z)]$, which figures so prominently in Eqs. (4.38) to (4.40), is closely related to the characteristic polynomial associated with the LP prediction filter; in fact, it has the same coefficients, appearing in the opposite order.

4.8. REFLECTION FOR SIGNALS OF KNOWN PHASE

LP has difficulties when the number of data points is small, because the order of the prediction filter is limited by the size of the data sample. Zhu and Bax pointed out [31] that it is possible to double the effective number of points by making use of two key facts: In the indirect dimensions the phase of the signals is known beforehand, and over short time periods the line widths can be approximated by a single value. For severely truncated data, the effect of having twice as many data points, almost for free, can be a very significant improvement.

Here is how the procedure works. Assume the initial time delay is equal to zero and the signals are all in phase. Let d_k be the measured value at time $k\Delta t$, where $k = 0, \ldots, M - 1$. Estimate the line width as L; even though the various components may actually decay at different rates, since M is small the variation in decays will not be significant over the course of the sample. The estimated decay can be canceled out by multiplying the data with a growing exponential. Set

$$d_k' = d_k e^{\pi L k \Delta t} \qquad (4.41)$$

Of course this will amplify the noise towards the end of the data sample, but provided that $M\Delta t$ is not much more than $1/\pi L$ the degradation should be acceptable. Now the $\mathbf{d}'$ signal is approximately equal to a sum of nondecaying sinusoids, so we have

$$d_k' = \sum_{j=1}^{m} A_j e^{2\pi i k \Delta t f_j} \qquad (4.42)$$

where A_j and f_j are the amplitude and frequency of the jth sinusoid. In this situation it is easy to extrapolate the data array to negative time values. The result is

$$d'_{-k} = \sum_{j=1}^{m} A_j e^{2\pi i(-k)\Delta t f_j} = d_k'^* \tag{4.43}$$

that is, the extrapolated value for d'_{-k} is just the complex conjugate of d'_k.

It is now possible to apply the LP analysis to a time series that is almost twice as long as the original, namely

$$d'_{-M+1}, \ldots, d'_{-1}, d'_0, d'_1, \ldots, d'_{M-1} \tag{4.44}$$

(Notice that the zero point is not reflected.) As a result, the LP prediction order can be twice as large, yielding much better results from extrapolation or parametric fitting. Of course, the line width and phase outputs from a parametric analysis must be adjusted to compensate for the fact that we have altered the decay rate and shifted the time origin, but this is easy to do.

This procedure can also be used when the initial time delay is equal to one half the sampling interval. We can describe this mathematically by saying that the index k takes on the values $1/2, 3/2, \ldots, M - 1/2$. The same analysis applies as before, and now the reflected time series is exactly twice as long as the original:

$$d'_{-M+1/2}, \ldots, d'_{-3/2}, d'_{-1/2}, d'_{1/2}, d'_{3/2}, \ldots, d'_{M-1/2} \tag{4.45}$$

Here every point is reflected.

4.9. PERFORMANCE CONSIDERATIONS

As with any computational procedure, it is important to consider the time required and the reliability of the results. In LP, most of the effort is devoted to determining the LP coefficients, which usually involves an SVD calculation. Singular value decomposition of an $(L \times m)$ matrix requires on the order of $m^2 L$ operations. Consequently LP is much more practical when the filter order m is small, and becomes much less so when the filter order is large.

The reliability of the results is strongly dependent on the S/N. A statistical study [32] has shown that LP provides reliable estimates of the frequencies of those peaks it identifies. The line width and amplitude estimates are not as accurate (and may exhibit a systematic bias), particularly for peaks that fall in a crowded region of the spectrum. Furthermore, the ability of LP to identify peaks diminishes with decreasing S/N, making high filter orders, perhaps as high as 10 to 100 times the number of peaks, essential. Due caution should be exercised when the S/N is below 10.

4.10. POLYNOMIAL FACTORIZATION

Since factoring the characteristic polynomial plays such an important role in LP, and since the polynomials that occur have rather high order and are therefore difficult to factor, it seems appropriate to include a factoring algorithm here. The one we will present is based on an algorithm published in reference 33. Similar procedures have been reported to factor successfully polynomials of order greater than 1000, but you should take care to perform plenty of testing before relying on any new factoring routine.

Part of the reason great care is required is the numerical inaccuracy of computer arithmetic. The large exponents that occur in polynomials make double-precision arithmetic a *sine qua non* for any order higher than about 20. And even then, round-off errors can produce surprising results.

To be concrete, let $P(z)$ be a complex polynomial of order n. Ideally we would like to find the values of the roots of P, but in practice, we can only determine approximations $r_1, \ldots, r_n$. Should we choose the r's so that $P(r_i)$ is as close as possible to zero for each i (finding the *roots* of P), or should we choose the values for which the product

$$P_r(z) = (z - r_1) \cdots (z - r_n) \tag{4.46}$$

has coefficients as close as possible to those of P (*factoring P*)? Mathematically the results ought to be the same, but in the presence of round-off errors they are not. Even worse, it may turn out that even though $P(z)$ was originally calculated by multiplying out

$$P(z) = (z - s_1) \cdots (z - s_n) \tag{4.47}$$

(for some set of values $s_1, \ldots, s_n$), the values of $P(r_i)$ may be closer to zero than the values of $P(s_i)$. For that matter, the coefficients of $P_r(z)$ may be closer to the coefficients of $P(z)$ than are the coefficients of

$$P_\pi(z) = (z - s_{\pi(1)}) \cdots (z - s_{\pi(n)}) \tag{4.48}$$

where π is a permutation of $\{1, \ldots, n\}$. That is, the value of the coefficients of the polynomial may depend on the order in which the monomials are multiplied together. Both of these phenomena occur quite often when dealing with orders above 50.

It is not always easy to tell which criterion for finding the roots is more appropriate. For LP extrapolation (when the characteristic polynomial is factored in order to regularize the roots and will be reconstituted to obtain the revised LP filter coefficients) it is probably best to use the factoring criterion.

The algorithm given here factors P, and so the roots it finds are not generally the points at which the value of P is closest to zero. It works by finding a root r_1 of P, then *deflating* P [i.e., dividing $P(z)$ by $(z - r_1)$] to get P^*, then finding a root r_2 of P^*, deflating P^* to get P^{**}, then finding a root r_3 of P^{**}, and so on. The operations involved in the synthetic divisions (the deflation steps) are almost exactly the inverse of the operations involved in synthetic multiplication by a monomial. Consequently, when the monomials $(z - r_n)$, $(z - r_{n-1})$, ..., $(z - r_1)$ are multiplied back together in the *reverse* order from the way they were found, the resulting polynomial has coefficients very close to those of P.

The deflection calculation is trivial, so we need only show how to find a root of a general polynomial. The calculation proceeds in two parts: initial use of Newton's method to approximate a root, and then application of Svejgaard's algorithm to pin down the value as precisely as possible. (The first part is not strictly necessary, but it does speed up the process.) Newton's method is well known: Given a guess z_k for the root, the next guess is taken to be

$$z_{k+1} = z_n + \Delta z, \qquad \text{where} \qquad \Delta z = -P(z_k)/P'(z_k) \qquad (4.49)$$

The starting guess z_0 can be set equal to zero. It is important to monitor $|P(z_k)|$ and $|P(z_{k+1})|$ during this process; if the absolute value fails to decrease then the next guess is no improvement over the current one. In this case Δz should be decreased by some small factor (say four) and z_{k+1} revised accordingly. Also, it is possible for the division by the derivative of P in Eq. (4.49) to overflow or underflow. We take all such difficulties (as well as the inability to find a good value for the next guess) as an indication that it is time to stop using Newton's method and proceed with the second part of the calculation. We never perform more than about 30 iterations of Newton's method in any event.

The second part uses a technique described as "the method of the rotating cross" to locate a local minimum of $|P(z)|$. At each there is a current guess z_k and an offset Δz. The polynomial is evaluated at the four points of a cross: The values $|P(z_k + \Delta z)|$, $|P(z_k + i\Delta z)|$, $|P(z_k - \Delta z)|$ and $|P(z_k - i\Delta z)|$ are computed and the smallest one is noted. This value is compared with $|P(z_k)|$. If it is smaller, then z_{k+1} is taken to be the corresponding point of the cross and Δz is multiplied by 1.5 (the cross is enlarged). Otherwise z_{k+1} is taken to be the current point z_k and Δz is multiplied by $(0.4 + 0.3i)$ (the cross is shrunk by a factor of two and rotated). This process continues until the cross is sufficiently small that $|z_k|$ and $|z_k| + |\Delta z|$ are indistinguishable to within the limits of double precision. The final value of z_k is then the root of P.

One might think that deflation affects the accuracy of the roots, since each deflation introduces round-off error into the coefficients, and that each root should be fine-tuned (or *polished*) using the original polynomial rather than the deflated polynomial. In practice, the errors are very small and the procedure works very well as it stands. In fact, polishing against P would detract from the accuracy of the factorization, although it might improve the accuracy of the roots.

4.11. AN EXAMPLE OF LP EXTRAPOLATION

We conclude with one final example of LP extrapolation, this time in the t_1 dimension of a two-dimensional data set. The results offer significant improvement over conventional processing (without extrapolation) and show why the use of LP for processing multidimensional data is becoming routine. Figure 4.5 contains one column extracted from data for a COSY experiment on a 16-residue peptide, before and after LP extrapolation. The additional points obtained by LP extrapolation allow the use of window functions that decay more slowly to zero, thereby yielding higher resolution, without introducing truncation artifacts. (Note that the window function will deemphasize the extrapolated points, reducing the effects of any errors in the LP extrapolation.)

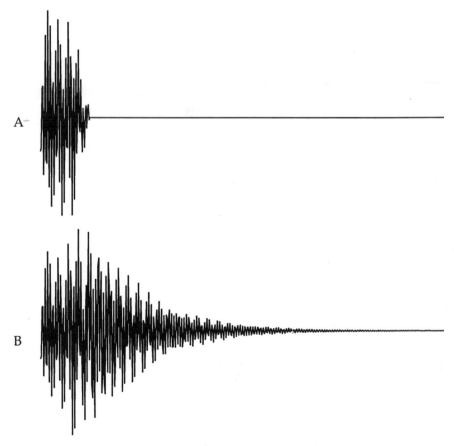

Fig. 4.5. LP extrapolation permits the use of much gentler apodization functions. Here we see a column (along t_1) of a two-dimensional data set, consisting of 64 points, following apodization (**A**). LP extrapolation to 512 points and apodization yields the signal in (**B**); the longer decay will result in higher resolution in the final spectrum.

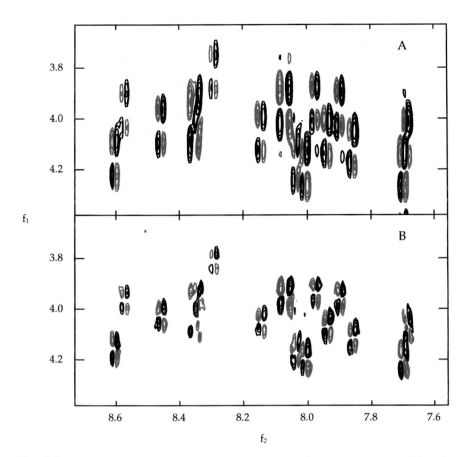

Fig. 4.6. Contour plots of a region from the two-dimensional spectrum containing the column shown in Figure 4.5, processed without (**A**) and with (**B**) LP extrapolation, illustrate the improved resolution afforded by LP. Note particularly the crowded region near 8.0 ppm in f_2, where it is easier to pick out individual multiplets.

Figure 4.6 compares contour plots of a portion of the spectrum computed without and with LP extrapolation.

TO READ FURTHER

An excellent review of time-domain fitting techniques, primarily LP, is given in reference 34.

5

MAXIMUM ENTROPY RECONSTRUCTION IN NMR: AN ALTERNATIVE TO DFT

We have seen that linear prediction provides a more accurate means of extrapolating an FID than does zero-filling, and that LP can be used to model the time series, as well as to extrapolate it. LP is not always effective, however. It is strictly correct only for exponentially decaying sinusoids; it requires relatively high S/N for the results to be reliable; and it is only applicable to time series that have been sampled at uniform intervals.

5.1. SPECTRUM ANALYSIS AS AN INVERSE PROBLEM

A more general approach, that avoids the limitations of the DFT and LP, treats spectrum analysis of a time series as an inverse problem. *Inverse problems* are those in which the properties of interest can only be viewed indirectly; a famous example is Plato's allegory of the cave [35], in which the problem is to infer the characteristics of an object from the shadows in the firelight it casts on the walls of a cave (we resist the temptation to inject a metaphor linking scientists to cave dwellers). In NMR spectrum analysis, the inverse problem is that of recovering the spectrum $\mathbf{f}$ of an ensemble of spins when the only observation we have is the FID, which is contaminated by noise. Symbolically, if $\hat{d}_k$ is the *true* nuclear resonance signal at time $k\Delta t$, d_k is the *measured* FID, and ε_k is the noise, then

$$d_k = \hat{d}_k + \varepsilon_k \qquad (5.1)$$

We can approach this inverse problem by placing restrictions on the possible solutions $\mathbf{f}$. The first restriction amounts to the obvious requirement that $\mathbf{f}$ must be consistent with the experimental observations. The agreement can be quantified through the use of a constraint statistic, determined by comparing a "mock" FID, $\mathbf{m} = \mathbf{ker} \cdot \text{IDFT}(\mathbf{f})$, with the actual data. (Here $\mathbf{ker}$ is a convolution kernel that is multiplied, point by point, with the inverse DFT of $\mathbf{f}$. We will discuss its use later; for now we will assume it consists of all ones.) When the noise and experimental error in the measured data are normally distributed, an appropriate statistic is the (unweighted) χ^2 value,

$$\chi^2(\mathbf{f}) = \sum_{k=0}^{M-1} |m_k - d_k|^2 \qquad (5.2)$$

where M is the number of data samples collected. Other forms of the constraint statistic $C(\mathbf{f})$ are possible, but χ^2 is the one generally used, because of its simplicity and the difficulty in justifying any other distribution of errors in the data. (A weighted χ^2 statistic can also be used—for example, if there is good reason to believe that some of the data points are more subject to error than others.) We will use $C(\mathbf{f}) = \chi^2(\mathbf{f})/2$, which will simplify some later calculations. The constraint that the reconstructed spectrum must be consistent with the measured data takes the form

$$C(\mathbf{f}) \leq C_0 \qquad (5.3)$$

where C_0 is an upper bound on the allowed error. Given a prior estimate of the amount of noise in the data, C_0 should be comparable to the power of the noise,

$$C_0 \approx \tfrac{1}{2} \sum_{k=0}^{M-1} |\varepsilon_k|^2 \qquad (5.4)$$

5.2. SOLVING THE INVERSE PROBLEM USING MAXIMUM ENTROPY RECONSTRUCTION

The constraint statistic alone does not provide enough of a restriction on the possible reconstructions to be of much use. In fact, any mock FID that matches the experimental FID to within C_0 for the first M points will satisfy the constraint. If we are to improve over the DFT, we need some additional criterion for regularizing the reconstructed spectrum, which will effectively constrain the values of the mock FID beyond M. One such criterion is the maximum entropy principle, which says that a reasonable reconstruction should add no new information beyond that contained in the experimental data. The principle originated in the work of Claude Shannon on the information-carrying capacity of circuits [36], where he showed that the entropy $S(p)$ of a probability distribution p, defined by

$$S(p) = -\sum_n p_n \log p_n \qquad (5.5)$$

is a measure of the lack of information expressed by p. There is a formal link between Shannon's entropy and the statistical entropy of an ensemble, and this link can be used to constrain possible spectrum reconstructions. According to the maximum entropy principle, the best spectrum reconstruction is the one having the highest entropy among all those consistent with the measured data. The process of finding this spectrum is called *maximum entropy reconstruction* (MaxEnt). [Maximum entropy reconstruction should be distinguished from the maximum entropy method (MEM), introduced by Burg, which is discussed in Chapter 4 as it is more closely related to LP.] Some people take the extreme position that the maximum entropy principle is the only philosophically justifiable and logically consistent regularizer to use [37]. On a more pragmatic level, it can be viewed simply as a means of obtaining a smooth reconstruction. Entropy is not unique as a regularizer, however, and there are other functionals that have similar properties.

An important distinction between MaxEnt reconstruction and methods of spectrum analysis (such as those based on LP) that implicitly assume Lorentzian line shapes, is that MaxEnt reconstruction *per se* makes no assumptions about the signal to be recovered. Yet it is flexible enough to allow prior information to be incorporated into the reconstruction. This can be done through the convolution kernel **ker**. For example, if the FID is known to decay at a certain rate, or to be J-modulated at a certain frequency, these features can be included in the kernel. The reconstruction then yields the *deconvolved* spectrum (i.e., one from which these features have been removed) having the highest entropy.

5.3. A NUMERICAL ALGORITHM

The maximum entropy principle supplies a formal basis for reconstructing the spectrum **f**, but some additional details must be addressed in order to apply MaxEnt reconstruction to NMR data. For modern NMR experiments, we need to be able to reconstruct complex spectra containing both positive and negative components. Any formulation for the entropy should be expressible in terms of a regularization function,

$$S(\mathbf{f}) = -\sum_{n=0}^{N-1} R(f_n) \tag{5.6}$$

where $R(f_n)$ is the negative of the contribution of a spectral component f_n to the overall entropy. The Shannon formula [Eq. (5.5)] has this form, but clearly does not apply to complex-valued spectra. A more suitable approach is to equate the entropy of a spectrum with the entropy of a corresponding ensemble of nuclear spins [38]. The derivation is rather involved, so we just present the result. For the so-called $S_{1/2}$ entropy, appropriate for particles with spin 1/2, the formula for R is

$$R(z) = \frac{|z|}{def} \log\left(\frac{|z|/def + \sqrt{4 + |z|^2/def^2}}{2}\right) - \sqrt{4 + |z|^2/def^2} \quad (5.7)$$

where *def* is a scale factor. Since $R(z)$ depends only on the absolute value of z, it is invariant under changes in phase, which we expect it should be. Although a value for *def* can be derived from the definition of $S_{1/2}$, the value turns out to be rather large. Such large values of *def* give rather poor reconstructions, resembling just the zero-filled DFT. More useful reconstructions are obtained if *def* is allowed to take smaller values, so we will treat it as an adjustable parameter.

The next detail that we are faced with is how to solve the constrained optimization problem, that is, maximizing $S(\mathbf{f})$ subject to the constraint that $C(\mathbf{f}) \leq C_0$. Since the entropy [Eqs. (5.6) and (5.7)] and the constraint statistic [Eq. (5.2)] are both everywhere convex, there is a unique, global solution. Furthermore, this maximum satisfies $C(\mathbf{f}) = C_0$, provided that the trivial solution $\mathbf{f} = \mathbf{0}$ does not also satisfy the constraint. Such constrained problems are often solved by converting them into an equivalent unconstrained optimization problem; for MaxEnt reconstruction, the equivalent problem is to maximize the objective function

$$Q(\mathbf{f}) = S(\mathbf{f}) - \lambda C(\mathbf{f}) \quad (5.8)$$

where λ is a Lagrange multiplier. The solution we seek corresponds to a critical point of Q, i.e., a point where $\nabla Q = \mathbf{0}$. There is no general analytic solution, and so we are forced to use numerical techniques. Quite often, numerical search using the gradient of the objective function (steepest ascent or conjugate gradients) is a reasonably efficient method for finding the critical point. Unfortunately, an objective function constructed from two such dissimilar functionals as the χ^2 statistic and the entropy is very difficult to maximize. Practical solution of the general MaxEnt reconstruction problem requires a fairly complex optimizer, and the most efficient algorithms require on the order of 100 or more times the computational effort of the DFT.

A particularly robust and efficient algorithm that paved the way for practical use of MaxEnt was invented by Skilling and Bryan [39]. This method uses multiple search directions and a variable metric to determine the size of the step of each iteration. The algorithm described below is a variant we have found useful in our laboratory.

The algorithm begins with a flat trial spectrum equal to zero everywhere. At each iteration, a mock FID is computed from the current value of the trial spectrum. This involves inverse Fourier transformation and multiplication by the convolution kernel, which is specified by the user as an input parameter. The first M points of the N-point mock FID are used to compute the value of $C(\mathbf{f})$. The algorithm constructs a small set of direction vectors, and computes a quadratic approximation to the entropy in the subspace spanned by these vectors. Since the constraint is itself quadratic, it is possible to maximize analytically the entropy approximation subject to the constraint in this subspace. This results in the trial spectrum for the next iteration.

The quadratic approximation to the entropy is accurate only over a small range, so when the value of $C(\mathbf{f})$ is far from the desired value C_0, it is better to solve the analytic maximization using a constraint value C_0' intermediate between C_0 and the current value of C, rather than attempt one large step. Consequently the algorithm proceeds in two phases. In the first phase, $C(\mathbf{f})$ is larger than C_0 and the algorithm attempts to find a spectrum having a lower value of C. In the second phase, C_0 has been attained and the algorithm seeks to maximize the entropy.

One possible direction vector is ∇S, the gradient of the entropy with respect to $\mathbf{f}$ [calculated by differentiating Eq. (5.6)]. Unfortunately, the gradient can be dominated by small values of f_n, and the algorithm would spend more time adjusting values that are close to zero than adjusting signal peaks. The key insight of Skilling and Bryan was that the introduction of an "entropy metric" results in an algorithm that more evenly weights the contributions of large and small values of the spectrum. Using the entropy metric amounts to replacing ∇S and ∇C (the gradient of the constraint with respect to the trial spectrum) with $\mathbf{H}^{-1}\nabla S$ and $\mathbf{H}^{-1}\nabla C$, where $\mathbf{H}$ is $-\nabla^2 S$, the negative of the Hessian (the matrix of second derivatives) of S. The matrix $\mathbf{H}$ is nearly diagonal, so its inverse is easy to compute. The search directions used during the first phase of the algorithm are:

$$\nabla C, \quad \mathbf{H}^{-1}\nabla C, \quad \mathbf{H}^{-1}\nabla Q, \quad \text{and} \quad \mathbf{H}^{-1}(\nabla^2 C)\mathbf{H}^{-1}\nabla Q \qquad (5.9)$$

where $Q = S - \lambda C$ and λ is set to $|\nabla S|/|\nabla C|$. Here $|\nabla S|$ and $|\nabla C|$ are the vector norms of ∇S and ∇C; the norm of a vector $\mathbf{v}$ is defined by

$$|\mathbf{v}| = \sqrt{\sum_{i=0}^{N-1} |v_i|^2} \qquad (5.10)$$

During the second phase, the second search direction is replaced by ∇Q. In the original algorithm of Skilling and Bryan, neither ∇C nor ∇Q are used as search directions, but we have found that including them speeds convergence. The major effort in computing these vectors stems from four discrete Fourier transformations (two each to compute ∇C and to apply $\nabla^2 C$) and multiplication by $\mathbf{H}^{-1}$.

With these direction vectors, a quadratic model for the entropy and the constraint can be constructed using a second-order Taylor expansion about the trial spectrum. A spectrum $\mathbf{f}'$ in the subspace spanned by the direction vectors can be represented by a vector $\mathbf{a}$ of length four: The correspondence is given by

$$\mathbf{f}' = a_1\mathbf{v}_1 + a_2\mathbf{v}_2 + a_3\mathbf{v}_3 + a_4\mathbf{v}_4 + \mathbf{f} \qquad (5.11)$$

where $\mathbf{v}_1$ through $\mathbf{v}_4$ are the direction vectors and $\mathbf{f}$ is the current trial spectrum. The quadratic approximations then become

$$S(\mathbf{a}) \approx S(\mathbf{f}) + \mathbf{B} \cdot \mathbf{a} + \tfrac{1}{2} \mathbf{a} \cdot \mathbf{\Lambda} \cdot \mathbf{a}$$

$$\text{and} \quad C(\mathbf{a}) \approx C(\mathbf{f}) + \mathbf{G} \cdot \mathbf{a} + \tfrac{1}{2} \mathbf{a} \cdot \mathbf{\Gamma} \cdot \mathbf{a}$$

(5.12)

where vectors $\mathbf{B}$ and $\mathbf{G}$ are inner products of ∇C and ∇S with the direction vectors:

$$B_i = \nabla S \cdot \mathbf{v}_i \quad \text{and} \quad C_i = \nabla C \cdot \mathbf{v}_i$$

(5.13)

and similarly, the (4×4) matrices $\mathbf{\Lambda}$ and $\mathbf{\Gamma}$ are the inner products of $\nabla^2 S$ and $\nabla^2 C$ with the direction vectors:

$$\Lambda_{ij} = \mathbf{v}_i \cdot \nabla^2 S \cdot \mathbf{v}_j \quad \text{and} \quad \Gamma_{ij} = \mathbf{v}_i \cdot \nabla^2 C \cdot \mathbf{v}_j$$

(5.14)

Computing these values requires several more Fourier transformations and large matrix products. The Lagrange condition $\nabla S - \lambda \nabla C = 0$ for the maximum of the objective function in the subspace becomes

$$(\mathbf{B} + \mathbf{\Lambda} \cdot \mathbf{a}) - \lambda(\mathbf{G} + \mathbf{\Gamma} \cdot \mathbf{a}) = 0$$

(5.15)

A change of coordinates in the subspace can simplify this equation by simultaneously diagonalizing $\mathbf{\Lambda}$ and $\mathbf{\Gamma}$. The solution is then given by

$$a_i = \frac{B_i - \lambda G_i}{\lambda \Gamma_{ii} - \Lambda_{ii}}$$

(5.16)

The value of λ can be found by using a binary search to determine the value for which $C(\mathbf{a})$ equals the target value C_0'.

Convergence criteria for determining when the trial spectrum is sufficiently close to the MaxEnt reconstruction derive from the Lagrange condition and the requirement that $C(\mathbf{f}) = C_0$. The Lagrange condition implies that ∇C and ∇S are parallel, so convergence can be monitored by computing the value

$$Test = \left| \frac{\nabla S}{|\nabla S|} - \frac{\nabla C}{|\nabla C|} \right|$$

(5.17)

The algorithm terminates when $C(\mathbf{f}) \approx C_0$ and $Test \ll 1$. In practice, we stop the algorithm when $Test < 10^{-3}$, or when a preset maximum number of iterations has been executed.

The value of C_0 can be estimated from the data by examining a portion of an FID that is essentially free of signal, or alternatively, by examining a signal-free region of the DFT spectrum. For MaxEnt reconstructions, C_0 should generally be somewhat larger than the estimate of the noise power. Much larger values of C_0 result in overly smooth reconstructions in which weaker components are washed out.

The other adjustable parameter, *def*, is rather more difficult to prescribe. *Def* represents the scale at which the nonlinear effects of MaxEnt reconstruction become significant. A large value of *def* results in nearly linear reconstructions; when *def* is large and C_0 is zero, the resulting reconstruction is essentially the same as the DFT of the zero-filled FID. On the other hand, very low values of *def* can give rise to slow convergence or numerical overflow. As a rule of thumb, we use values of *def* somewhat lower than the noise level in the spectrum. Fortunately, the results are not overly sensitive to the choice of *def*. We don't know of any algorithmic procedure for determining the best value, but theoretical considerations suggest that $def = \sqrt{C_0/MN}$ is a reasonable choice.

5.4. CALCULATION OF GRADIENTS AND HESSIANS

Up to now we've talked about gradients and Hessians of C, S, and Q, without showing explicitly how to calcualte them. They are sufficiently complicated and important for us to examine them in some detail. (We will also see in Chapter 6 that the form of the gradients illuminates the relationship between MaxEnt and some other methods of spectrum reconstruction.)

In computing the various gradients, there is a subtle point to consider: Since **f** is complex, it is necessary to compute the partial derivatives with respect to both the real and imaginary components. We will combine the two partial derivatives into a single complex value, so that by definition,

$$\frac{\partial C}{\partial f_n} = \frac{\partial C}{\partial f_n^r} + i \frac{\partial C}{\partial f_n^i}, \tag{5.18}$$

where the superscripts r and i denote the real and imaginary components, respectively. A similar consideration applies to the Hessian matrices. Note that this implies that the inner products in Eqs. (5.13) and (5.14) are not the usual inner products of complex vectors. Instead, $\nabla C \cdot \mathbf{v}$ must be interpreted as

$$\sum_{n=0}^{N-1} \nabla C_n^r v_n^r + \nabla C_n^i v_n^i \tag{5.19}$$

In essence ∇C and **v** are treated as real vectors of length $2N$, rather than complex vectors of length N. Of course, the other products must also be handled in this way.

We start with the gradient of the constraint. Combining the definitions of $C(\mathbf{f})$ and **m,** the constraint is given by

$$C(\mathbf{f}) = \tfrac{1}{2} \sum_{k=0}^{M-1} |ker_k \text{ IDFT}(\mathbf{f})_k - d_k|^2 \tag{5.20}$$

where **ker** is the convolution kernel. We have already seen that the IDFT can be represented by a unitary matrix $\mathbf{F}^\dagger$. The kernel can also be represented in matrix form, by taking $\mathbf{K}$ to be an $(M \times N)$ matrix with diagonal elements $K_{kk} = ker_k$. Then we can write

$$C(\mathbf{f}) = \tfrac{1}{2} |\mathbf{KF}^\dagger \mathbf{f} - \mathbf{d}|^2 \tag{5.21}$$

This can be further simplified by setting $\mathbf{A}$ equal to the matrix $\mathbf{KF}^\dagger$, so $C(\mathbf{f}) = \tfrac{1}{2} |\mathbf{Af} - \mathbf{d}|^2$. Allow us one last substitution: set $\mathbf{b} = \mathbf{Af} - \mathbf{d}$. The elements of $\mathbf{b}$ are given by

$$b_k = \left[\sum_{j=0}^{N-1} (A_{kj}^r f_j^r - A_{kj}^i f_j^i) - d_k^r \right] + i \left[\sum_{j=0}^{N-1} (A_{kj}^r f_j^i + A_{kj}^i f_j^r) - d_k^i \right] \tag{5.22}$$

The constraint is equal to

$$C(\mathbf{f}) = \tfrac{1}{2} \sum_{k=0}^{M-1} |b_k|^2 = \tfrac{1}{2} \sum_{k=0}^{M-1} (b_k^r)^2 + (b_k^i)^2 \tag{5.23}$$

and using the chain rule,

$$\frac{\partial C}{\partial f_n^r} = \sum_{k=0}^{M-1} b_k^r \frac{\partial b_k^r}{\partial f_n^r} + b_k^i \frac{\partial b_k^i}{\partial f_n^r} = \sum_{k=0}^{M-1} b_k^r A_{kn}^r + b_k^i A_{kn}^i \tag{5.24}$$

and

$$\frac{\partial C}{\partial f_n^i} = \sum_{k=0}^{M-1} b_k^r \frac{\partial b_k^r}{\partial f_n^i} + b_k^i \frac{\partial b_k^i}{\partial f_n^i} = \sum_{k=0}^{M-1} b_k^r (-A_{kn}^i) + b_k^i A_{kn}^r \tag{5.25}$$

The full gradient is

$$\frac{\partial C}{\partial f_n} = \sum_{k=0}^{M-1} (A_{kn}^r b_k^r + A_{kn}^i b_k^i) + i(A_{kn}^r b_k^i - A_{kn}^i b_k^r)$$
$$= \sum_{k=0}^{M-1} (A_{kn})^* b_k = (\mathbf{A}^\dagger \mathbf{b})_n \tag{5.26}$$

Thus

$$\nabla C = \mathbf{A}^\dagger \mathbf{b} = \mathbf{A}^\dagger (\mathbf{Af} - \mathbf{d})$$
$$= \mathbf{FK}^\dagger (\mathbf{KF}^\dagger \mathbf{f} - \mathbf{d}) \tag{5.27}$$

Multiplying an M-element vector by the matrix $\mathbf{K}^{\dagger}$ corresponds to multiplying point by point with the convolution kernel and zero-filling from M to N elements.

To calculate the Hessian of C, expand Eq. (5.27):

$$\nabla C = (\mathbf{A}^{\dagger}\mathbf{A})\mathbf{f} - \mathbf{A}^{\dagger}\mathbf{d}$$

$$\frac{\partial C}{\partial f_m^r} = \sum_{n=0}^{N-1} [(A^{\dagger}A)_{mn}^r f_n^r - (A^{\dagger}A)_{mn}^i f_n^i] - (A^{\dagger}d)_m^r \qquad (5.28)$$

$$\frac{\partial C}{\partial f_m^i} = \sum_{n=0}^{N-1} [(A^{\dagger}A)_{mn}^r f_n^i + (A^{\dagger}A)_{mn}^i f_n^r] - (A^{\dagger}d)_m^i$$

Taking second derivatives yields

$$\frac{\partial^2 C}{\partial f_m^r \partial f_n^r} = (A^{\dagger}A)_{mn}^r \qquad \frac{\partial^2 C}{\partial f_m^r \partial f_n^i} = -(A^{\dagger}A)_{mn}^i$$

$$\frac{\partial^2 C}{\partial f_m^i \partial f_n^r} = (A^{\dagger}A)_{mn}^i \qquad \frac{\partial^2 C}{\partial f_m^i \partial f_n^i} = (A^{\dagger}A)_{mn}^r \qquad (5.29)$$

So when $\nabla^2 C$ multiplies a vector $\mathbf{v}$, the result is

$$(\nabla^2 C \cdot \mathbf{v})_m = \sum_{n=0}^{N-1} \left[\frac{\partial^2 C}{\partial f_m^r \partial f_n^r} v_n^r + \frac{\partial^2 C}{\partial f_m^r \partial f_n^i} v_n^i \right] + i\left[\frac{\partial^2 C}{\partial f_m^i \partial f_n^r} v_n^r + \frac{\partial^2 C}{\partial f_m^i \partial f_n^i} v_n^i \right]$$

$$= \sum_{n=0}^{N-1} [(A^{\dagger}A)_{mn}^r v_n^r - (A^{\dagger}A)_{mn}^i v_n^i] + i[(A^{\dagger}A)_{mn}^i v_n^r + (A^{\dagger}A)_{mn}^r v_n^i]$$

$$= \sum_{n=0}^{N-1} (A^{\dagger}A)_{mn} v_n = (\mathbf{A}^{\dagger}\mathbf{A}\mathbf{v})_m \qquad (5.30)$$

In this sense we can simply say that

$$\nabla^2 C = \mathbf{A}^{\dagger}\mathbf{A} = \mathbf{F}\mathbf{K}^{\dagger}\mathbf{K}\mathbf{F}^{\dagger} \qquad (5.31)$$

In other words, applying $\nabla^2 C$ is the same as performing an IDFT, truncating to M points, multiplying by the convolution kernel *twice*, zero-filling back to N points, and forward transforming.

The gradient of S does not have a comparably tidy matrix representation, but it is simpler by virtue of the fact that the nth element of the gradient depends only on the nth component of $\mathbf{f}$. The principal difficulty is keeping track of the real and imaginary parts. We can spare ourselves from drowning in a sea of superscripts and subscripts by introducing $z = |f_n|$, where $f_n = x + iy$ and $z = \sqrt{x^2 + y^2}$. The real and imaginary components of the gradient of S are

$$-\frac{\partial R(z)}{\partial x} = -R'(z)\frac{x}{z} \quad \text{and} \quad -\frac{\partial R(z)}{\partial y} = -R'(z)\frac{y}{z} \tag{5.32}$$

where $R'(z)$ is the derivative of R with respect to z. For the $S_{1/2}$ entropy [Eq. (5.6)],

$$R'(z) = \frac{1}{def}\log\left(\frac{z/def + \sqrt{4 + z^2/def^2}}{2}\right) \tag{5.33}$$

Combining Eqs. (5.32) and (5.33), the total gradient is

$$\frac{\partial S(\mathbf{f})}{\partial f_n} = \frac{-f_n}{def|f_n|}\log\left(\frac{|f_n|/def + \sqrt{4 + |f_n|^2/def^2}}{2}\right) \tag{5.34}$$

The jk component of the Hessian vanishes unless $j = k$, so we may continue to dispense with subscripts. However, the effect of the dependence on the absolute value is to mix the real and imaginary parts, so we cannot ignore the mixed partial derivatives. The second derivatives of R are

$$\frac{\partial^2 R(z)}{\partial x^2} = R''(z)\frac{x^2}{z^2} + R'(z)\left(\frac{1}{z} - \frac{x^2}{z^3}\right)$$

$$\frac{\partial^2 R(z)}{\partial y^2} = R''(z)\frac{y^2}{z^2} + R'(z)\left(\frac{1}{z} - \frac{y^2}{z^3}\right) \tag{5.35}$$

$$\frac{\partial^2 R(z)}{\partial x \partial y} = \frac{\partial^2 R(z)}{\partial y \partial x} = R''(z)\frac{xy}{z^2} + R'(z)\left(\frac{-xy}{z^3}\right)$$

The matrix of second derivatives can be written as a simple sum of (2×2) matrices,

$$\left(R''(z) - \frac{R'(z)}{z}\right)\begin{bmatrix} x^2/z^2 & xy/z^2 \\ xy/z^2 & y^2/z^2 \end{bmatrix} + \frac{R'(z)}{z}\begin{bmatrix} 1 & 0 \\ 0 & 1 \end{bmatrix} \tag{5.36}$$

$R''(z)$ is obtained by differentiating Eq. (5.33),

$$R''(z) = \frac{1}{def^2}\frac{1}{\sqrt{4 + z^2/def^2}} \tag{5.37}$$

Combining Eqs. (5.36) and (5.37) yields the total Hessian,

$$\nabla^2 S(z) = \left\{ \frac{-1}{def^2} \frac{1}{\sqrt{4 + z^2/def^2}} \right.$$

$$+ \frac{1}{z \, def} \log\left(\frac{z/def + \sqrt{4 + z^2/def^2}}{2} \right) \right\}$$ (5.38)

$$\begin{bmatrix} x^2/z^2 & xy/z^2 \\ xy/z^2 & y^2/z^2 \end{bmatrix}$$

$$- \frac{1}{z \, def} \log\left(\frac{z/def + \sqrt{4 + z^2/def^2}}{2} \right) \begin{bmatrix} 1 & 0 \\ 0 & 1 \end{bmatrix}.$$

Just as the complex gradient can be viewed as a length $2N$ real vector, the complex Hessian can be viewed as a $(2N \times 2N)$ real matrix. The matrix is block diagonal, the blocks are size (2×2), and Eq. (5.38) gives the value of each block.

5.5. ANALYTIC SOLUTION

While numerical solution is required in the general case, there is a special case of MaxEnt reconstruction that has an analytical solution. Although it is unrealistic in several respects, examining it can give some insights into how MaxEnt reconstruction works. It arises when N (the number of points in the reconstructed spectrum) is equal to M (the number of experimental data points), and when the relationship between the trial spectrum and the mock FID is given simply by the inverse Fourier transform (that is, we are not trying to deconvolve a kernel). Under these circumstances, Parseval's theorem permits the constraint statistic to be computed in the frequency domain, and the Lagrange condition becomes

$$\mathbf{0} = \nabla Q = \nabla S - \lambda(\mathbf{f} - \mathbf{g})$$ (5.39)

where $\mathbf{g}$ is the DFT of the data $\mathbf{d}$. The solution is given by

$$|f_n| = \delta_\lambda^{-1}(|g_n|) \quad \text{and} \quad \text{phase}(f_n) = \text{phase}(g_n)$$ (5.40)

where $\delta_\lambda^{-1}(x)$ is the inverse of the function

$$\delta_\lambda(z) = z + R'(z)/\lambda$$ (5.41)

This result corresponds to a nonlinear transformation, applied point by point to the DFT of the time domain data. The transformation depends on the value of λ, and has the effect of scaling every point in the spectrum down, but points closer to the baseline are scaled down more than points far above the baseline.

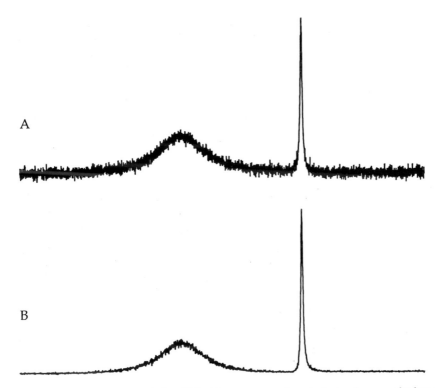

Fig. 5.1. A typical characteristic of MaxEnt reconstructions is that noise near the baseline is suppressed more than noise superimposed on top of broad peaks. (**A**) is a DFT spectrum and (**B**) is the corresponding MaxEnt reconstruction.

This helps to explain a characteristic of MaxEnt reconstructions: Noise near the baseline is suppressed more than noise away from the baseline (for example, superimposed on a broad peak) (Fig. 5.1). Figure 5.2 shows $\delta_\lambda^{-1}(x)$ for various values of λ. As λ increases, the weight given to the constraint term in the objective function increases, and the transformation approaches the identity function.

5.6. SENSITIVITY VERSUS SIGNAL-TO-NOISE RATIO

Sensitivity is the ability to distinguish signal from noise. S/N is often used as an indicator of sensitivity, and for linear methods such as apodization, an improvement in S/N implies an improvement in sensitivity. However, for nonlinear methods such as MaxEnt reconstruction, this is not so.

We just saw a special case of MaxEnt reconstruction in which the reconstructed spectrum is equivalent to the result of applying a nonlinear transformation to each point of the DFT spectrum. Since the same transformation is

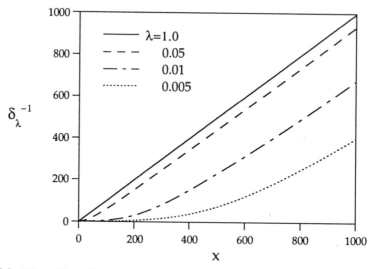

Fig. 5.2. When $N = M$ and there is no convolution kernel, MaxEnt reconstruction reduces to the application of a nonlinear transformation $\delta_\lambda^{-1}(x)$ to the DFT spectrum. As λ grows larger, placing more weight on the constraint relative to the entropy, the transformation becomes closer to the identity function. The effect of $\delta_\lambda^{-1}(x)$ is to scale down the valule of x; smaller values are scaled down more than larger ones.

applied to both the signal and to the noise, peaks that are comparable in height to the noise level will be reduced by the same amount as the noise. The ratio between the highest signal peaks and the noise may increase, but small peaks will be just as difficult to distinguish as before. This means that the sensitivity is unchanged, and the improvement in S/N is purely cosmetic.

It does not follow that gains in S/N are also purely cosmetic in the more general case where $M \neq N$ or a convolution kernel is applied. To assess sensitivity in these settings, some measure other than S/N must be used; one method is described in Chapter 7. In spite of some extravagant claims, the gains in sensitivity using MaxEnt that have been convincingly demonstrated are only comparable to those that could be achieved using the matched filter and conventional processing [40]. A prudent investigator will always question whether gains in S/N really correspond to gains in sensitivity.

5.7. DECONVOLUTION

What is the point of the convolution kernel **ker** that we have mentioned on several occasions? It commonly happens that the envelope of the FID results in unwanted manifestations in the spectrum. For example, a rapidly decaying envelope gives rise to broad lines; or modulation of the FID at a fixed frequency, such as multiplication by $\cos(\pi Jt)$, will cause each peak to split into a pair of peaks. Now according to the convolution theorem, the result of applying any envelope to the FID is to convolve the spectrum with the DFT of the envelope. The process of removing these unwanted manifestations from a spectrum is called *deconvolution*.

The kernel **ker** is simply the envelope whose effect we want to remove. The most straightforward way to deconvolve the kernel would be to divide d_k by ker_k for each k. Unfortunately, the kernels that occur in NMR usually pass close to zero, so the division is numerically unstable and tends to greatly enhance the noise. MaxEnt, on the other hand, can perform numerically stable deconvolution. Since the algorithm attempts to match the mock FID with the experimental data, and the mock FID is constructed by multiplying the inverse DFT of the trial spectrum by the kernel, the trial spectrum itself ends up lacking the effects of the kernel. Of course, you must know beforehand the form of the kernel, but we will see later that this knowledge does not have to be exact.

5.8. EXAMPLE APPLICATIONS

One application of MaxEnt is to reduce truncation artifacts without losing resolution. In Figure 5.3, the same data set has been processed in four different ways: zero-filled DFT, windowed zero-filled DFT, LP-extrapolated windowed DFT, and MaxEnt reconstruction. With unwindowed DFT processing (Fig. 5.3A), the region around 1.5 ppm shows artifacts due to truncation. Windowing the FID (Fig. 5.3B) reduces the artifacts, at the expense of lower resolution due to broadening; more aggressive resolution enhancement would improve the resolution at the expense of more pronounced negative wings surrounding the tall peaks. Extrapolation by LP prior to windowing and DFT (Fig. 5.3C) reduces the artifacts somewhat, without broadening. The MaxEnt reconstruction (Fig. 5.3D) is virtually free of artifacts and has narrow lines; further line narrowing could be achieved by deconvolving the natural line width, as we will see below.

Figure 5.4 illustrates gains in S/N that can be achieved using MaxEnt reconstruction. Figure 5.4A and B contain DFT spectra of the same sample, but the data for Figure 5.4B were collected using a 7° observe pulse to reduce the S/N. Figure 5.4C shows the MaxEnt reconstruction of the data from Figure 5.4B. It bears repeating that, although striking, the improvement in S/N does not necessarily correspond to a real gain in sensitivity. Note, for example, that the multiplet structure of the peak near 3.25 ppm is not much more evident in Figure 5.4C than in Figure 5.4B.

A characteristic of MaxEnt reconstruction that distinguishes it from parametric methods, such as LP, is that it makes no implicit assumptions about the nature of the data (although using MaxEnt reconstruction to perform deconvolution does presume the convolution kernel to be correct). MaxEnt is thus less susceptible to bias when the signal has unusual characteristics (such as markedly nonexponential decay), when the S/N is low, or when the data have

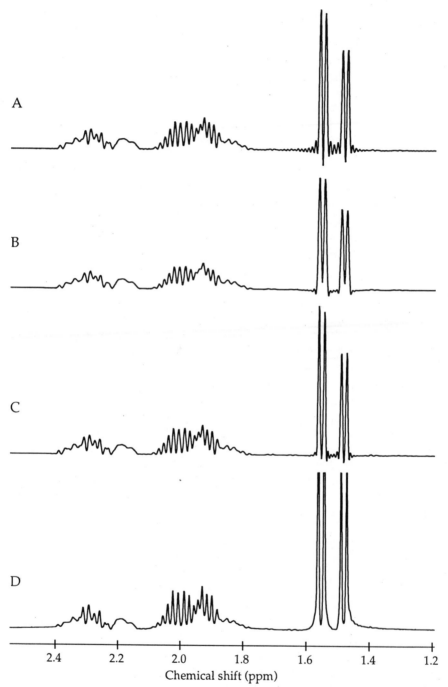

Fig. 5.3. MaxEnt has the ability to reduce truncation artifacts without broadening peaks. Compare the results of processing a 1024-point FID to obtain an 8192-point spectrum: (**A**) Using zero-filling and DFT; (**B**) using 60°-shifted sine bell apodization, zero-filling, and DFT; (**C**) using LP extrapolation to 2048 points (filter order 100), 60°-shifted sine bell apodization, zero-filling, and DFT; and (**D**) using MaxEnt reconstruction.

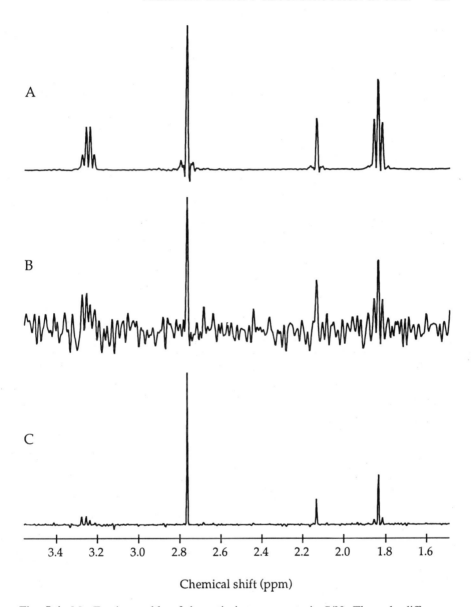

Fig. 5.4. MaxEnt is capable of dramatic improvements in S/N. The only difference between (**A**) and (**B**) is that the data for (**B**) were collected using a 7° observe pulse to reduce the S/N. Both data sets were processed using 60°-shifted sine bell apodization, zero-filling from 1024 to 4096 points, and DFT. In (**C**) the data for (**B**) were processed using MaxEnt reconstruction. The ratio of the peak heights to the noise level (the S/N) is vastly improved compared to (**B**), but that does not mean that the ability to distinguish weak peaks (the sensitivity) is any better.

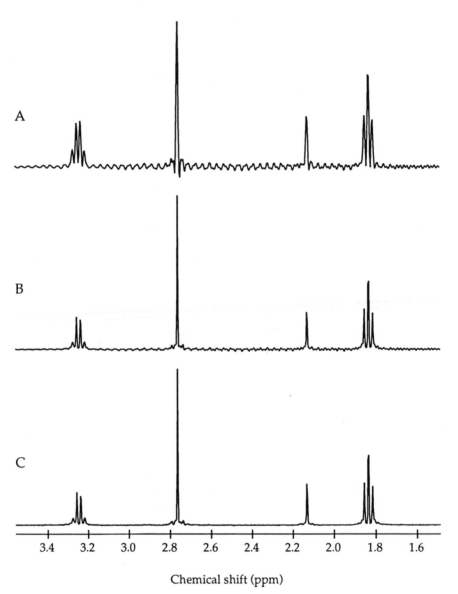

Chemical shift (ppm)

Fig. 5.5. MaxEnt is able to handle corrupted data. Here is a contrived example, constructed by manually corrupting the one-dimensional FID used to generate Figure 5.4A: The points from 350 to 359 and from 750 to 759 (out of a total of 1024 points) were set to zero. (**A**) shows the DFT spectrum of the corrupted FID, zero-filled to 4096 points. (**B**) and (**C**) are 4096-point MaxEnt reconstructions. In (**B**), all 1024 points of the FID were included in the computation of $C(\mathbf{f})$, while in (**C**), the zeroed points were left out. These results illustrate that MaxEnt is much more robust with regard to data contamination than the DFT, particularly when the locations of the corrupted data points are known.

been corrupted. It is quite common for points at the beginning of an FID to contain errors, which can result in distorted baselines. In these circumstances, backwards LP can be used to correct the corrupted points. If the erroneous data points are in the middle of an FID, however, or are distributed in some unusual fashion throughout the FID, LP is less useful. MaxEnt reconstruction, on the other hand, does not require the data to be sampled at uniformly spaced intervals; the missing data points can simply be omitted from the calculation of $C(\mathbf{f})$. This means that MaxEnt can be applied quite easily to data that are corrupted in an irregular fashion. It is our own sad experience that points in the indirect dimensions of long multidimensional experiments are occasionally contaminated by intermittent instrument failure or environmental influences. MaxEnt reconstruction can recover useful spectra from such data. An example is shown in Figure 5.5.

We described earlier how MaxEnt reconstruction can be used to deconvolve the effect of the envelope of the FID on the spectrum. A typical application is to enhance resolution by deconvolving the effect of an exponential decay. Figure 5.6 includes two MaxEnt reconstructions of the same FID. In Figure 5.6A no convolution kernel was applied, and in Figure 5.6B the kernel consisted of an exponential decay corresponding to a 4-Hz line width. Although the peaks are narrower (and better resolved) in Figure 5.6B, the reduction in line width is smaller than 4 Hz. We will see later that the optimal convolution kernel generally does not correspond to the natural line width.

MaxEnt reconstruction can also be used to deconvolve an arbitrary modulation, such as J-modulation. This can greatly simplify a crowded spectrum consisting of multiplets, as long as they all have the same structure (i.e., in-phase or antiphase) and approximately the same splitting, since the convolution kernel applies indiscriminately to all the peaks. In Figure 5.7A, MaxEnt reconstruction was used to process the t_2 dimension of a two-dimensional experiment without deconvolution; in Figure 5.7B deconvolution of a 7-Hz J-coupling was incorporated. The effect is that each in-phase doublet has been replaced by a single peak at the center frequency. The deconvolution is not perfect: Note the "ghost" peak in Figure 5.7B at $f_1 = 21$ ppm, $f_2 = 0.88$ ppm.

5.9. CHOOSING THE PARAMETERS

The choice of the MaxEnt reconstruction parameters has a significant effect on the appearance of the final spectrum. The parameters are not independent, so it is important to optimize them all together. Figure 5.8 shows how C_0 affects the reconstruction. If it is too large (Fig. 5.8A and B) the weak features will be lost, and the reconstruction is overly broad. On the other hand, if C_0 is too small (Fig. 5.8D), the reconstruction is forced to match the FID closely, even including the noise. Figure 5.8E shows the unapodized, zero-filled DFT spectrum, for comparison. Note that the truncation artifacts, so evident in Figure 5.8E, are not present in the MaxEnt reconstructions.

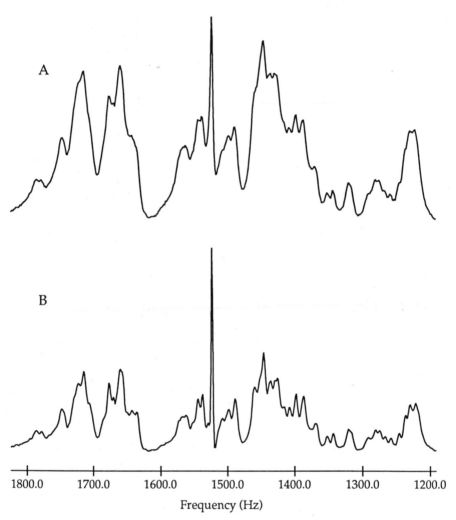

Fig. 5.6. MaxEnt can be used to deconvolve a decay kernel, improving the resolution of the spectrum. In (**A**), MaxEnt reconstruction was performed without deconvolving a kernel. In (**B**), a kernel corresponding to a 4-Hz exponential decay was deconvolved, using otherwise identical parameters for the reconstruction. Notice that the lines are sharper and the noise level has not increased.

Figure 5.9 shows reconstructions using the same data set but different values of *def*. When *def* is large in comparison to the peak heights (Fig. 5.9A), the reconstructions resemble the zero-filled DFT, including truncation artifacts. The explanation can be found by examining the behavior of the $S_{1/2}$ entropy [Eqs. (5.6) and (5.7)] as *def* becomes very large; $S(\mathbf{f})$ becomes proportional to $-\Sigma |f_n|^2/def^2$ plus a constant, so MaxEnt reconstruction reduces to minimum power reconstruction. Since the power is the same in the time and frequency

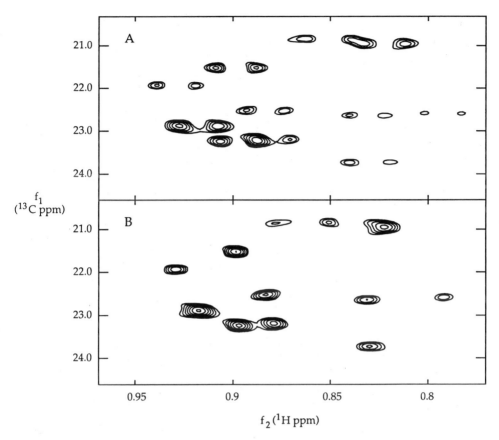

Fig. 5.7. MaxEnt can be used to deconvolve J-modulation. In (**A**), MaxEnt reconstruction was used in the f_2 dimension. (**B**) is the same, except that a convolution kernel consisting of a cosine with frequency 3.5 Hz (corresponding to a 7-Hz coupling) was applied. As a result, the J-doublets have been collapsed to singlets—the splitting has been removed.

domains, the result is equivalent to minimum power extrapolation, which is just zero-filling! As the value of *def* is decreased, the nonlinearity of MaxEnt reconstruction becomes more evident. In Figure 5.9B, the value of *def* is comparable to the noise level, and the noise remains visible in the reconstruction. In Figure 5.9C, the value of *def* is smaller than the noise level, and the noise level in the reconstruction is greatly reduced. Had there been low-intensity peaks in the spectrum, comparable to the noise level, they would have been reduced in intensity as well. The nonlinearity of MaxEnt can be seen in Figure 5.9C by examining the relative heights of the components of the quintet near the center of the spectrum.

The interaction of the parameters C_0 and *def* is illustrated in Figure 5.10. The reconstructions in Figure 5.10A and B use a value for C_0 that is smaller

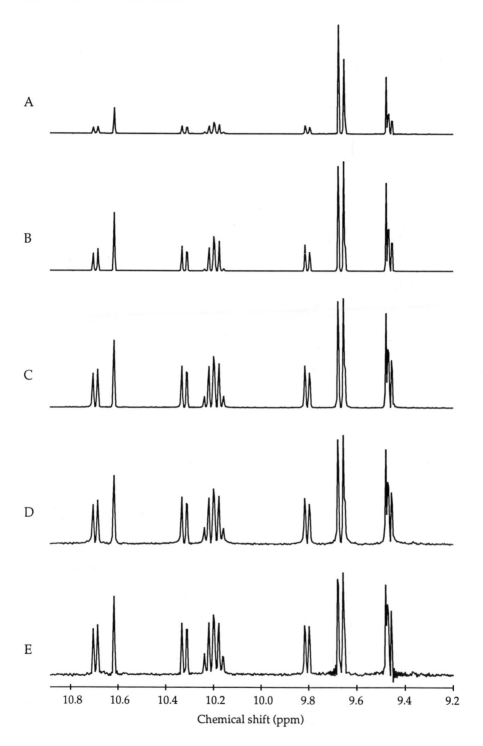

Chemical shift (ppm)

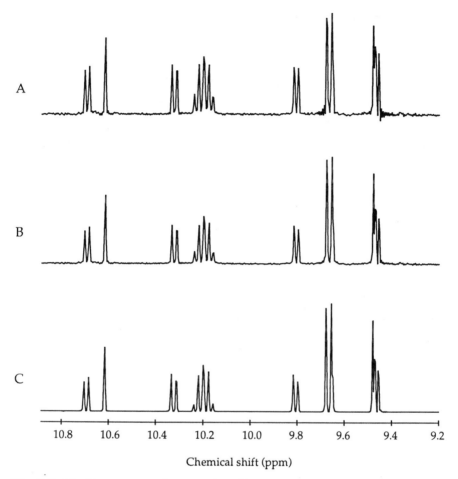

Fig. 5.9. MaxEnt reconstructions are also affected by the value of *def*, although not so dramatically as by C_0. These three reconstructions used *def* equal to (**A**) 1000, (**B**) 100, and (**C**) 1. Large values of *def* lead to reconstructions resembling the DFT spectrum; smaller values lead to more nonlinear reconstructions that are better for suppressing noise but that also distort relative peak intensities.

←

Fig. 5.8. MaxEnt reconstructions are strongly affected by the choice of the parameters. Here the value of C_0 (which approximates the noise level in the signal) has been set to (**A**) 170, (**B**) 110, (**C**) 50, and (**D**) 1. Large values cause weak features of the spectrum to be lost, whereas small values cause even the minute details of the noise to be included. (**E**) is the nonapodized zero-filled DFT spectrum.

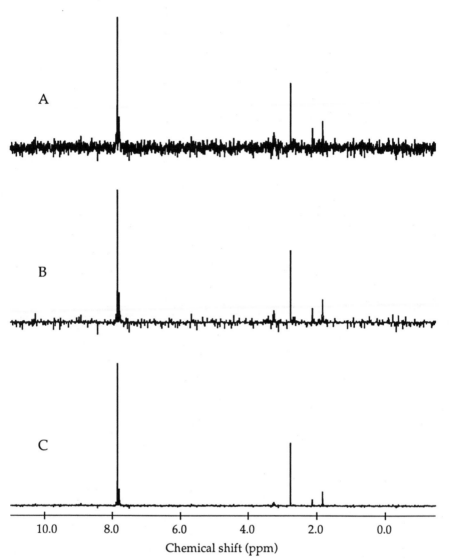

Fig. 5.10. Certain combinations of *def* and C_0 can lead to undesirable results in MaxEnt reconstruction. Here the values of *def* and C_0 are (**A**) 10, 1.5; (**B**) 0.001, 1.5; and (**C**) 0.1, 2.8, respectively. In (**B**), the nonlinear effects of MaxEnt reconstruction are especially pronounced over the range of the noise, leading to a spiky appearance.

than the noise power and different values for *def*. For this C_0, the nonlinear effects resulting from small values of *def* become significant over the range of the noise level. Consequently parts of the noise are suppressed more than others, leading to a spiky appearance (Fig. 5.10B). The reconstruction in Figure 5.10C uses the same *def* as Figure 5.10B, but C_0 has been increased, so that the noise is suppressed much more evenly.

Figures 5.8–5.10 should make it clear how to achieve visually stunning MaxEnt reconstructions: Choose C_0 sufficiently large to suppress all the features you don't like and *def* sufficiently small to smooth the points between the surviving features. The results may not always be scientifically valid, but they will impress your friends. Obtaining useful reconstructions requires a more careful selection of the parameters, and a modicum of error analysis.

The choice of convolution kernel also influences the result of MaxEnt reconstruction. Figure 5.11 shows a series of MaxEnt reconstructions deconvolving exponential decays corresponding to line widths ranging from 0 Hz

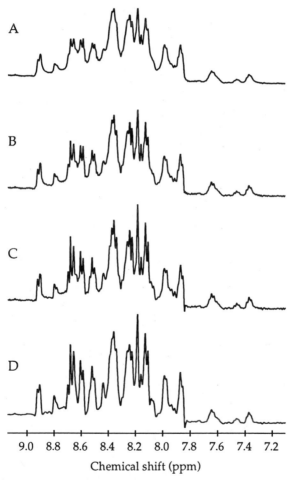

Fig. 5.11. The best choice for the decay rate for the convolution kernel in MaxEnt reconstruction is usually smaller than the natural line width of the peaks in the spectrum. The values of the decay rate used here correspond to Lorentzian line widths of (**A**) 0 Hz (effectively no deconvolution), (**B**) 2 Hz, (**C**) 4 Hz, (**D**) 6 Hz. The best spectrum is the one in (**C**), in spite of the fact that the intrinsic line widths are close to 8 Hz.

(effectively no deconvolution) to 6 Hz. The best resolution is obtained using a decay that corresponds to a line width of 4 Hz, which is less than the 8-Hz natural line width. Using too small or too large a decay results in broader spectra; the negative wings that begin to appear when using a decay that is too large are reminiscent of the wings that occur when DM or CD apodization is used overly aggressively to improve resolution.

You may wonder why MaxEnt reconstruction doesn't yield extremely narrow lines when the width of the convolution kernel matches the natural line width. There are two reasons: One is that the lines are not perfectly Lorentzian, so that a synthetic exponential decay does not exactly match the actual decay. The other is that the entropy functional conspires against the intense lines that would result; broadening them (to the extent possible without $C(\mathbf{f})$ exceeding C_0) reduces the intensity at the central frequency and so increases the entropy. Extreme narrowing of lines by MaxEnt reconstruction is thus dependent on a delicate counterbalance of the effects of the entropy functional and the constraint statistic. Even in instances where it is possible to narrow the lines below the digital resolution, leakage will occur when the frequency is not a multiple of $1/N\Delta t$, just as for the DFT.

5.10. MAXENT RECONSTRUCTION IN MULTIPLE DIMENSIONS

MaxEnt reconstruction can be applied to multidimensional data, and it is here that the ability to obtain high resolution spectra from short data records offers truly significant benefits. Since the experiment time increases linearly with the amount of data collected in the indirect dimensions of a multidimensional data set, resolution in these dimensions is often limited by the amount of time available. An advantage of MaxEnt reconstruction is that the data samples do not need to be collected at uniform intervals, as we discuss below in the section on nonlinear sampling.

There is a number of ways MaxEnt reconstruction can be extended to multiple dimensions. The most general is simply to compute the entire spectrum in one go. This can present a formidable computational challenge, however, since each iteration of the reconstruction algorithm requires several transformations of the entire data set from the frequency domain to the time domain and back. An alternative that is far less demanding is to apply MaxEnt reconstruction to each vector along a particular dimension independently. Using this approach, MaxEnt reconstructions of large multidimensional data sets can be performed efficiently on typical workstations. With a small adjustment to the algorithm, a series of one-dimensional reconstructions yields results that are nearly equivalent to the more general multidimensional reconstruction.

The necessary modification of the algorithm is easy to explain. In one-dimensional MaxEnt reconstruction, one seeks to maximize the objective function with λ chosen so that $C(\mathbf{f}) = C_0$ is satisfied. When computing MaxEnt reconstructions for the separate vectors of a multidimensional data set, normal variations in the noise and the signal can result in different values for λ for each vector. The consequences of this variation are often insignificant, but they can be important if they are large enough to perturb the line shapes, or if quantification of the reconstructed spectrum is performed. Since the nonlinearity of MaxEnt reconstruction depends on the relative weights applied to the entropy and the constraint statistic, it is important to ensure that the same weight (i.e., value of λ) applies to each vector. This can be accomplished by choosing a typical vector, estimating the noise, and computing a normal one-dimensional MaxEnt reconstruction. Each vector in the multidimensional data set is then reconstructed, using the value of λ obtained from the reconstruction of the trial vector instead of iterating to achieve $C(\mathbf{f}) = C_0$.

For the more general approach, in which the entropy and the constraint are evaluated for the entire spectrum at once, the calculations are more computationally demanding, but there are no additional conceptual difficulties. The entropy and constraint do not directly depend on the number of dimensions; they merely involve sums over all the elements. There is one new wrinkle, however: When dealing with hypercomplex spectra, the absolute values that appear in the formula for R [Eq. (5.7)] must be treated as the square root of the sum of the squares of *all* the hypercomplex components. The formulas for ∇S and $\nabla^2 S$ can readily be generalized.

5.11. NONLINEAR SAMPLING

MaxEnt reconstruction has the advantage over the DFT and LP that it can handle data that have been sampled at arbitrary times, rather than at uniform intervals. This is called *nonlinear sampling*, and it offers the benefit of improved resolution or sensitivity when the number of sample points is limited. In theory, the samples could be chosen at completely arbitrary times. A more useful guiding principle is similar to the idea of a matched filter: Acquire more samples at times when the signal is more intense, and fewer samples when it is less intense. To keep things manageable, we will retain the notion of a uniformly spaced sequence of times separated by intervals Δt, but we will only acquire data samples at some of the times. The *sampling schedule* is the subset of times that are actually sampled; an M out of N point sampling schedule consists of a selection of M values from the full linear schedule $\{0, \Delta t, 2\Delta t, \ldots, (N - 1)\Delta t\}$. For a given number of samples M, using only a subset of the full linear schedule means that the last data sample can be taken later than $(M - 1)\Delta t$, which would be the last sample time for normal linear sampling. The total duration of an experiment is more or less independent of the number of samples in the acquisition dimension, so the benefits of nonlinear sampling accrue mainly in the indirect dimensions.

It has been almost axiomatic up to this point that fewer samples means poorer resolution, since the digital resolution is given by $1/N\Delta t$. However, what really matters is not just the number of samples, but also the time of the last data point. Nonlinear sampling allows us to collect data out to a time $N\Delta t$, while only collecting M samples. There are three obvious ways that we can use this freedom to our advantage. One way is simply to improve resolution without collecting more samples, and so without lengthening the experiment time. Another way is to shorten the experiment duration, without sacrificing resolution. Finally, the extra time can be used to perform additional signal averaging, thereby improving sensitivity.

Optimal schedules are devised from consideration of the envelope of the FID. Three classes can be distinguished according to the shape of the envelope, depending on whether the signal decays exponentially (e.g. NOESY, Fig. 5.12A), is a sine-modulated exponential decay (e.g. COSY, Fig. 5.12B), or does not decay (constant-time experiments, Fig. 5.12C). A sampling schedule to match the envelopes in Figure 5.12A or B can be determined by the following procedure. We start by defining a continuous function that represents the desired density of samples. For an exponential decay corresponding to a line width L, the function takes the form

$$D(t) = Ae^{-\pi Lt} \tag{5.42}$$

The density function is normalized, so that

$$\int_0^{N\Delta t} D(t)\, dt = M - 1 \tag{5.43}$$

where M is the number of exponentially sampled points to be distributed over a time interval spanned by N uniform periods Δt. The parameter L is set to match the actual decay rate of the FID. Solving Eq. (5.43) yields

$$A = \frac{(M - 1)\pi L}{1 - e^{-\pi LN\Delta t}} \tag{5.44}$$

A sine-modulated signal requires a density function of the form

$$D(t) = Ae^{-\pi Lt} \sin(\pi t/N\Delta t) \tag{5.45}$$

(Here we are assuming that the duration of the sampling, $N\Delta t$, is equal to one half-period of the sine modulation; the treatment can be generalized.) Substituting the density function (5.45) into Eq. (5.43) leads to

$$A = \frac{(M - 1)\,(\pi L^2 + (1/N\Delta t)^2)(N\Delta t)}{1 + e^{-\pi LN\Delta t}} \tag{5.46}$$

We then calculate the cumulative density function

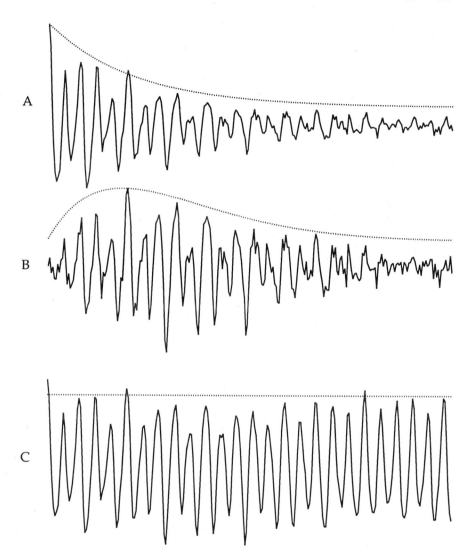

Fig. 5.12. Sampling schedules for nonlinear sampling should be tailored to match the envelope of the signal. Three common types of envelopes occur in NMR: (**A**) exponential decay; (**B**) sine-modulated exponential decay; and (**C**) no decay. The envelope of each signal is indicated by the dotted line.

$$I(T) = \int_0^T D(t) \, dt \qquad (5.47)$$

for $T = \Delta t, \ldots, N\Delta t$. For each integer j from 0 to $M - 1$, the delay time t_j for the jth data point is taken to be the least integer multiple of Δt greater than t_{j-1} such that $I(t_j) \geq j$. The result is an M out of N point sampling schedule;

there are M samples, the spacing of the samples mirrors the density function, and the last sample is at time $N\Delta t$.

Figure 5.13 shows examples of optimal sampling schedules for exponential and sine-modulated exponential signal envelopes (Fig. 5.13A and B). The procedure described above doesn't work so well for signals that don't decay— the resulting schedules would consist simply of uniformly spaced times, but with a fixed interval that is larger than Δt. This is just a linear sampling schedule with reduced spectral width. It turns out that for nondecaying signals, choosing samples randomly from the full linear schedule works quite well (Fig. 5.13C) [41].

When applied to an indirect dimension of a multidimensional experiment, a reduction in the number of sample points translates into a proportional decrease in the total experiment time. Since multidimensional experiments typically require many hours or days, even a modest reduction in the experiment time can be important. Figure 5.14 shows that this is often possible using nonlinear sampling, without significant losses in resolution or sensitivity. For reference, Figure 5.14A shows a contour plot of data processed using LP extrapolation,

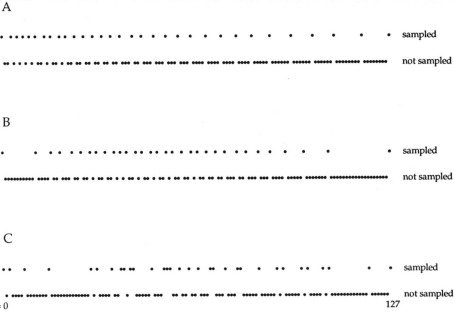

Fig. 5.13. Shown here are schematic depictions of sampling schedules appropriate for the three types of signal envelopes in Figure 5.12. Each contains 32 out of 128 points; the times to be sampled are indicated by dots in the upper line, and the times not sampled by dots in the lower line. For the exponential decay (**A**) and sine-modulated exponential decay (**B**), you can see that the density of samples matches the intensity of the envelope. For nondecaying signals, the envelope does not provide a basis for choosing sample points, but a randomized schedule (**C**) works reasonably well.

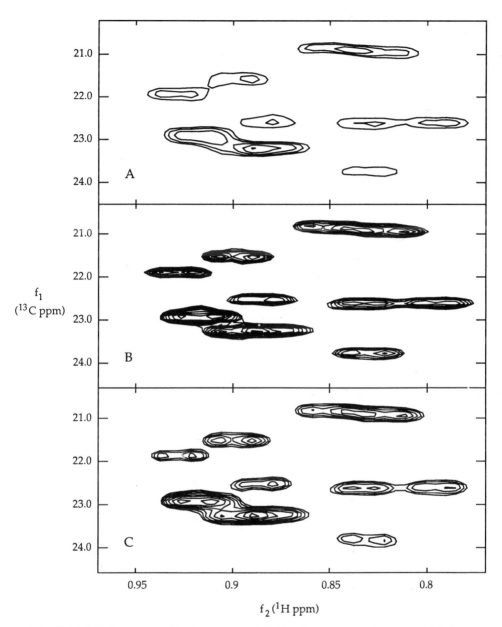

Fig. 5.14. Nonlinear sampling in concert with MaxEnt reconstruction can yield significant time savings in the collection of multidimensional data. Here is a region extracted from three two-dimensional spectra, all of the same sample. All three were processed using windowing and DFT in t_2. In (**A**), 128 points were collected in t_1 and processed using LP extrapolation to 1024 points, windowing, and DFT. In (**B**), an exponential sampling schedule with 128 out of 512 points was used in t_1, together with MaxEnt reconstruction deconvolving a 20-Hz Lorentzian line width. (**C**) is the same as (**B**), except that the sampling schedule consisted of only 64 out of 256 points. The time savings by a factor of two in (**C**) was achieved with little loss in sensitivity or resolution, particularly compared with conventional processing.

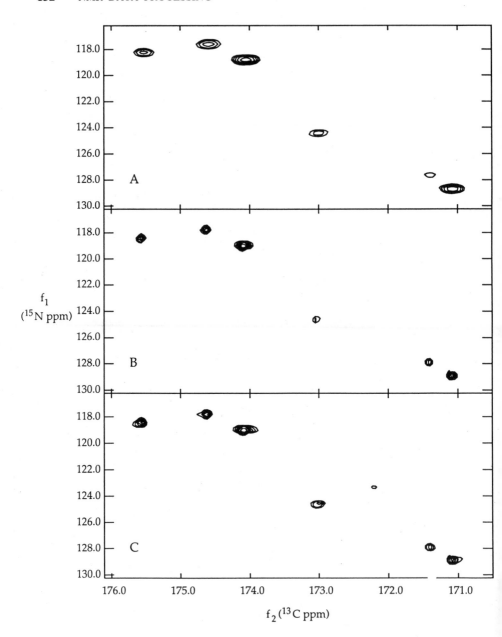

f_1
(^{15}N ppm)

f_2 (^{13}C ppm)

Fig. 5.15. Nonlinear sampling permits one to trade off the number of points collected in different dimensions with minimal loss of sensitivity or resolution. These cross sections from three-dimensional constant-time spectra were collected and processed using different combinations of linear and nonlinear sampling. (**A**) used linear sampling in all dimensions and conventional DFT processing; the number of points is 42 in the t_1 dimension and 48 in the t_2 dimension, zero-filled to 128 and 256, respectively; the time required to collect the data was 26 hours. (**B**) shows the results obtained using

windowing, and DFT in t_1. Figure 5.14B and C used nonlinear sampling in t_1, with different sampling schedules, and MaxEnt reconstruction. In Figure 5.14B, the same number of points was collected as in Figure 5.14A. The spectrum in Figure 5.14C used half as many points, requiring half the time to collect, but still has better S/N and resolution than the DFT spectrum.

One of the neat things about nonlinear sampling is that it gives you considerable flexibility to balance the competing requirements of S/N, resolution, and experiment time. Figure 5.15 illustrates the possibilities. It shows two-dimensional cross sections from three three-dimensional spectra. Figure 5.15A used linear sampling in all dimensions and conventional DFT processing. In Figure 5.15B, nonlinear sampling was used to reduce the number of points collected in the t_1 dimension, but some of the time savings was used to increase the number of points collected in the t_2 dimension. In Figure 5.15C, nonlinear sampling was used in *both* indirect dimensions. (Since t_1 and t_2 in this experiment were constant-time dimensions, the sampling schedules were random ones.) The results of the trade-off are apparent: Though it required less experiment time, the spectrum in Figure 5.15B has better resolution than the one in Figure 15.15A. The spectrum in Figure 5.15C required even less time, but the line widths in f_2 are broader and the peak shapes are slightly distorted. Nevertheless, the quality is still intermediate between that of Figure 5.15A and B.

MaxEnt reconstruction of nonlinearly sampled data is not without limitations. For a fixed number of samples, it can yield improved resolution, but at the cost of reduced sensitivity (for decaying signals). This is because some of the samples are collected at later times, when the signal intensity is lower. Furthermore, the processing can introduce additional noise. In essence, MaxEnt is being asked to fill in the sample points that are missing from the time-domain data; obviously it will not always be correct. The errors that arise will show up as artifacts in the reconstructed spectrum. These artifacts are usually much smaller than the largest peaks in the spectrum, but if the dynamic range is high, there is a danger of confusing weaker peaks with artifacts. (When strong diagonal peaks are present, one technique that can help reduce the artifacts is to model the diagonal components in the time-domain signal and subtract the model from the data prior to MaxEnt processing.) The nature of the artifacts will depend on the sampling schedule; further discussion can be found in reference 41.

---

random sampling to collect 16 out of 42 points in t_1 and increasing the number of points collected in t_2 to 90, requiring 19 hours. MaxEnt reconstruction was used in the t_1 dimension only. (**C**) shows results obtained using nonlinear sampling in *both* indirect dimensions; 16 out of 42 points were collected in t_1 and 45 out of 90 points were collected in t_2, requiring only 9.5 hours. Comparison of (**A**) and (**B**) shows that nonlinear sampling can improve resolution while keeping the total experiment time approximately constant. It can also be used to shorten the time required to collect useful data, as shown in (**C**).

5.12. QUANTIFICATION

The nonlinearity of MaxEnt reconstruction is an inherent characteristic; it is responsible for the method's ability to suppress noise without sacrificing resolution. Nonlinearity has important implications in situations where quantification of peak intensities or volumes is required, such as nuclear Overhauser effect measurements or difference spectroscopy. For difference spectroscopy, it may be sufficient to compute the difference using the time-domain data and compute the MaxEnt reconstruction of the difference. This assures that the same nonlinearity applies to both experiments. When measuring nuclear Overhauser effects, or otherwise quantifying peak intensities, there are two possible approaches. One is to tightly constrain the reconstruction to match the data, i.e., choose a small value for C_0, which for a given value of *def* forces the reconstruction to be more nearly linear (although at the expense of noise suppression). Another is to add synthetic signals of known intensity to the time-domain data prior to reconstruction. A calibration curve can then be constructed by quantifying the intensities of the known signals. An example of this approach is given in Chapter 7.

5.13. PERFORMANCE

A statistical study of the accuracy and bias of MaxEnt reconstruction has been published in reference 42. The authors concluded that estimates of peak intensities from MaxEnt reconstructions are comparable to the results obtained using least-squares fitting of a model line shape to the DFT spectrum. The more prior information available, the better the results obtained using MaxEnt reconstruction. For example, a convolution kernel derived from the line shape of an actual peak yields better results than a kernel consisting of a pure exponential decay.

The computational cost of MaxEnt reconstruction for one-dimensional spectra is typically two orders of magnitude greater than the cost of computing a DFT. In contrast to model-based methods such as LP, the computational effort of MaxEnt does not depend on the number of components in the signal. It *does* depend on the values of the parameters C_0, *def*, and **ker**; poor choices can dramatically slow the algorithm's convergence. For large multidimensional reconstructions, the effort is dominated by memory requirements, rather than the number of computations. With the algorithm described above, the required intermediate storage is on the order of 16 times the memory needed for the reconstructed spectrum. Although row-wise reconstruction can be used to ameliorate this requirement, fully multidimensional MaxEnt reconstructions remain the province of supercomputer-class machines with large amounts of high-speed random-access memory.

5.14. SUMMARY

MaxEnt reconstruction is a powerful method for computing spectral estimates that avoids some of the limitations inherent in the discrete Fourier transform. It is more general than methods that are restricted to uniformly sampled Lorentzian signals, yet it is capable of incorporating prior knowledge about the signal when it is available. The computational cost, while significantly greater than the DFT, is in a range that is practical on modern workstations, even for large data sets. Some of the power of the method stems from its nonlinearity, which dictates that the method must be used cautiously. Its ability to use nonlinearly sampled data makes it a valuable tool for improving the resolution and sensitivity of multidimensional experiments, or just for saving time. As the cost of state-of-the-art NMR spectrometers continues to soar, we expect that methods like MaxEnt will become increasingly important as a way of getting the most out of the investment.

TO READ FURTHER

The maximum entropy principle has been applied to data analysis in a wide variety of fields. The edited volumes of the proceedings of the annual maximum entropy conference give an overview [43–45].

6

EMERGING METHODS

In this chapter we describe three additional approaches to spectrum analysis of time series. We refer to them as "emerging" methods only in the sense that they have not been as widely used in NMR as the other methods described so far. The first method, iterated soft thresholding, has an especially simple implementation, but offers some surprising connections to MaxEnt and other methods. The second method uses the discrete wavelet transform, which like the DFT is based on expansion in a series of orthonormal basis vectors, but the basis vectors have some decidedly unusual properties. Wavelets are a fairly recent development, and applications in signal processing and spectrum analysis are still being discovered. The third method is a parametric approach called Bayesian analysis, which estimates the probability that a particular model accurately describes the signal. This sampling is by no means exhaustive, but simply represents a few of the methods we think hold some promise for application in NMR.

6.1. ITERATED SOFT THRESHOLDING

The key to MaxEnt reconstruction is the identification of a regularization functional that provides a measure of the quality of a spectrum. Out of all possible spectra that are consistent with the measured data, the one for which the regularization functional takes on the smallest value is selected as the best reconstruction. MaxEnt reconstruction uses the entropy functional, of course (actually, the negative of the entropy), but there are other possible functionals that might be used to obtain improved spectra. One possibility is the L_1 norm (the L_1 norm of a vector is just the sum of the absolute values of its members).

It has been shown that the method of "minimum L_1 projection" is capable of *perfect* reconstruction of band-limited signals contaminated by certain kinds of noise [46]. This result holds if the noise has "sparse support," that is, it is nonzero only for some small fraction of the points in the spectrum. Unfortunately, in NMR the noise does not have sparse support, but contaminates the spectrum everywhere. However, it is reasonable to presume that the minimum L_1 projection will still remain a good estimate of the spectrum if the S/N is not too low. We will show that *iterated soft thresholding* (IST) is a method for finding the minimum L_1 projection.

Iterated soft thresholding can be introduced through an intuitive line of reasoning; the formal description will come later. Suppose you have recorded an FID that contains several decaying sinusoids with similar amplitudes, and noise. Suppose further that the signal has been recorded for a time that is short compared to the decay rate of the sinusoids, and that you want to obtain a high-resolution spectrum. An estimate of the high-resolution spectrum can be obtained by zero-filling the FID prior to Fourier transformation; the DFT spectrum will contain truncation artifacts, of course (sinc wiggles). But now try the following: Choose a threshold that is smaller than the peak heights, but larger than the height of the truncation artifacts, and apply *soft thresholding*, i.e., set every point below the threshold in absolute value to zero, and subtract the value of threshold from each of the other points. The inverse DFT of the thresholded spectrum will not agree very well with the collected data, but compared to the zero-filled FID, it will do a better job of extending the data; not much better, but better. The process can be repeated: Replace the first part of the mock FID with the collected data, Fourier transform, and apply the threshold once again (Figure 6.1). Keep repeating this process until there is no significant change in the spectrum, that is, until a *fixed point* has been reached. In symbolic form, apply the operation

$$\mathbf{f}^{l+1} = T(\mathbf{f}^l) \tag{6.1}$$

where T represents replacement followed by soft thresholding, and $\mathbf{f}^l$ is the *l*th trial spectrum.

We call this method "soft" thresholding to distinguish it from "hard" thresholding, in which values below the threshold are set to zero but the others are unchanged. Papoulis has described an iterated hard thresholding algorithm in reference 47. The distinction may seem minor, but it permits an analysis of the method that makes clear its similarity to MaxEnt reconstruction, as we shall see.

The results of IST can appear pretty impressive, even after a relatively small number of iterations. Figure 6.2 compares a zero-filled DFT spectrum (Fig. 6.2A) with the result of applying IST for 20 iterations (Fig. 6.2B). Unfortunately, the spectrum in Figure 6.2B is not a fixed point, since the change caused by the last iteration was not negligible. Figure 6.2C shows the result of 200 iterations; it is noticeably different from Figure 6.2B, proving the al-

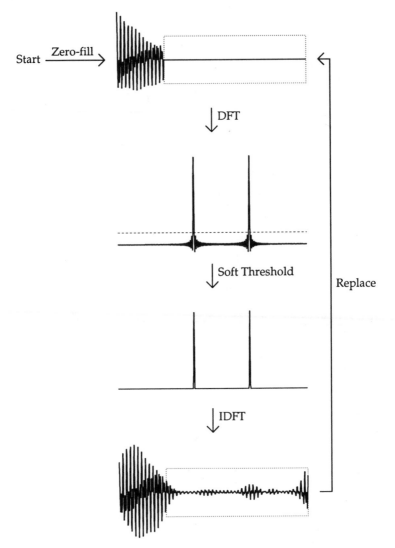

Fig. 6.1. A simple algorithm for computing the minimum L_1 norm projection consists of iterated soft thresholding. The initial trial spectrum is the DFT of the zero-filled data. Subsequent trial spectra are computed by applying the soft thresholding operation: setting points below a threshold to zero, and subtracting the threshold from all other points. An inverse Fourier transformation is applied to the result, the ''tail'' is used to extend the measured data, and the augmented data is Fourier transformed.

gorithm had not converged to a fixed point. To get a better feeling for what the algorithm is doing and what is special about the fixed point, we need to formalize the description of IST. In the process we will see how to generalize the algorithm to accommodate deconvolution and nonlinear sampling.

   The key to understanding IST is the L_1 norm. It turns out that the fixed point

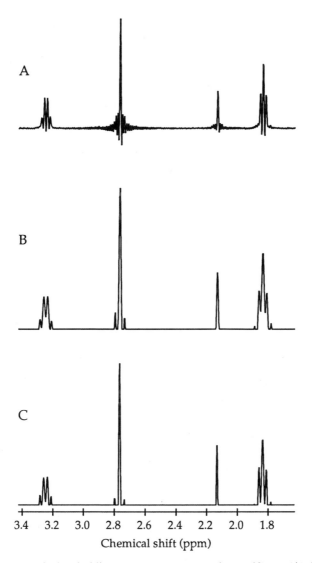

Fig. 6.2. Iterated soft thresholding can remove truncation artifacts. (**A**) is the zero-filled DFT spectrum. (**B**) and (**C**) show the results of 20 and 200 iterations of IST, respectively. A difficult problem is knowing when to stop. Although the truncation artifacts are gone after 20 iterations, the algorithm has not reached a fixed point.

of IST is simply a minimum L_1 norm reconstruction. Let's see why this is so. Not surprisingly, minimum L_1 norm reconstruction is highly analogous to MaxEnt reconstruction (discussed in Chapter 5): Find the spectrum **f** with minimum L_1 norm (instead of maximum entropy), subject to the constraint that the mock FID obtained from **f** is consistent with the data. Adopting the notation used in Chapter 5, find **f** so as to minimize

$$L_1(\mathbf{f}) = \sum_{n=0}^{N-1} R(f_n) \quad \text{where} \quad R(z) = |z| \tag{6.2}$$

subject to the constraint that

$$C(\mathbf{f}) = \tfrac{1}{2} \sum_{k=0}^{M-1} |ker_k \cdot \text{IDFT}(\mathbf{f})_k - d_k|^2 \le C_0 \tag{6.3}$$

As before, we convert the constrained optimization into an unconstrained optimization problem by introducing the objective function

$$Q(\mathbf{f}) = \tau L_1(\mathbf{f}) + C(\mathbf{f}) \tag{6.4}$$

where τ is a Lagrange multiplier. This formula is analogous to Eq. (5.8), with two changes. One is that the Lagrange multiplier is used to weight the regularization functional, rather than the constraint statistic—a trivial difference. The other is that the regularization functional is the L_1 norm, rather than the negative of the entropy. This difference is nontrivial. Notice, however, that the two functionals bear a certain resemblance. The negative entropy function [Eq. (5.7)] is roughly of the form $|z| \log |z|$, which is after all similar to $|z|$.

The solution spectrum $\mathbf{f}$ is the one that minimizes Q, so

$$\nabla Q(\mathbf{f}) = \tau \nabla L_1(\mathbf{f}) + \nabla C(\mathbf{f}) = \mathbf{0} \tag{6.5}$$

The functional $C(\mathbf{f})$ has the same form as in MaxEnt reconstruction, so ∇C is given by Eq. (5.26). That is, the gradient is obtained by computing the difference between the mock FID and the data, weighting by the convolution kernel, and applying a Fourier transform to the result. Compare this with the replacement operation in IST, in which the first part of the mock FID is replaced by the collected data, and the result is Fourier transformed. Amazingly enough, when no kernel is applied, this replacement operation is equivalent to moving along the gradient of C! This can be seen by expressing the result of the replacement operation as:

$$\text{DFT}[\text{IDFT}(\mathbf{f}) - trunc(\text{IDFT}(\mathbf{f})) + zerofill(\mathbf{d})] \tag{6.6}$$

where *trunc* is the operation of setting elements from M to N equal to zero, and *zerofill* is the operation of extending a vector from M to N elements with zeros. In essence, this formula says that the portion of the mock FID corresponding to the measured data is set to zero by subtracting $trunc(\text{IDFT}(\mathbf{f}))$ from $\text{IDFT}(\mathbf{f})$, and the resulting void is filled with the data $\mathbf{d}$. We can rewrite this as

$$\text{DFT}\{\text{IDFT}(\mathbf{f}) + zerofill[\mathbf{d} - shrink(\text{IDFT}(\mathbf{f}))]\} \tag{6.7}$$
$$= \mathbf{f} + \text{DFT}\{zerofill[\mathbf{d} - shrink(\text{IDFT}(\mathbf{f}))]\}$$

where *shrink* is the operation of throwing away the elements from $M + 1$ to N. This agrees with expression (6.6) because $trunc(\mathbf{x}) = zerofill(shrink(\mathbf{x}))$ for any vector $\mathbf{x}$. But now the second term is just $-\nabla C$, so the final result is equivalent to $\mathbf{f} - \nabla C$.

The gradient of R is more straightforward to compute. Expanding R, we have

$$R(\mathbf{f}) = \sum_{n=0}^{N-1} |f_n| = \sum_{n=0}^{N-1} [(f_n^r)^2 + (f_n^i)^2]^{1/2} \tag{6.8}$$

The components of the gradient are

$$\frac{\partial R}{\partial f_k} = [(f_k^r)^2 + (f_k^i)^2]^{-1/2} f_k^r + i[(f_k^r)^2 + (f_k^i)^2]^{-1/2} f_k^i = \frac{f_k}{|f_k|} \tag{6.9}$$

Subtracting $\tau f_k/|f_k|$ from f_k is equivalent to soft thresholding with threshold τ (at least, provided that $|f_k| \geq \tau$). So combining the replacement operation and the thresholding operation, we see that one iteration of IST corresponds to motion opposite the gradient of Q, and a fixed point of IST corresponds to a minimum of Q.

Now that we know that IST minimizes Q, we are in a better position to explore its convergence properties. In a manner analogous to that used in MaxEnt reconstruction, we will define *Test* as

$$Test = \left| \frac{\nabla R}{|\nabla R|} - \frac{\nabla C}{|\nabla C|} \right| \tag{6.10}$$

and we can introduce the step size, defined as

$$Stepsize = \left(\frac{1}{N} \sum_{n=0}^{N-1} |f_n^{l+1} - f_n^l|^2 \right)^{1/2} \tag{6.11}$$

Figure 6.3 shows Q, *Test*, and *Stepsize* as a function of l, the number of iterations. Since IST moves along ∇Q, the value of Q decreases monotonically, in contrast to *Test* and *Stepsize*. Figure 6.3 also shows that merely monitoring *Stepsize* is not a very safe way to test convergence, since it becomes small even while Q is still changing. *Test* gives a better criterion.

Now it's time to "'fess up." The astute reader will have noticed that there are some problems. First, ∇R is not defined for $|f_k| = 0$, and soft thresholding is not the same as subtracting $\tau f_k/|f_k|$ if $|f_k| < \tau$. Second, ∇R and ∇C are not evaluated at the same point: ∇C is evaluated for $\mathbf{f}^l$ at the beginning of the lth iteration, while ∇R is evaluated after the step along ∇C has been made. So the formal description we just laid out isn't quite the whole story (but it can be made more complete, at the expense of simplicity). Nevertheless, it does ex-

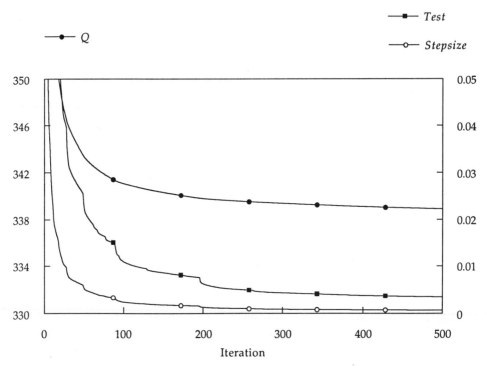

Fig. 6.3. Testing an iterated algorithm for convergence can be a significant challenge. Here the progress of IST is monitored by plotting the values of the objective function Q, *Test* (which is proportional to $|\nabla Q|$), and *Stepsize*. Note that Q and *Test* continue to decrease even after *Stepsize* has approached zero.

plain why the result of IST is the minimum L_1 norm projection, and it shows the relationship to MaxEnt reconstruction.

The formal description also indicates how IST can be generalized to include a convolution kernel and nonlinear sampling. All that is needed is to change the replacement operation so that it continues to coincide with motion along $-\nabla C$. The new operation amounts to computing the mock FID, subtracting it from **d**, multiplying by the convolution kernel (and setting to zero the points not in the sampling schedule), applying a DFT, and adding the result to **f**. You can check that this reduces to plain IST when there is no kernel and normal sampling is used.

Iterated soft thresholding offers some practical advantages over MaxEnt. One is that it requires only one parameter, the threshold level. Another is that the IST algorithm is particularly easy to implement. A drawback is slow convergence; this is especially troublesome when the spectrum to be reconstucted has high dynamic range. Figure 6.4A shows the results of IST applied to data in which the residual solvent resonance is about 10,000 times more intense than the solute resonances (and is off-scale in the figure). IST was iterated for

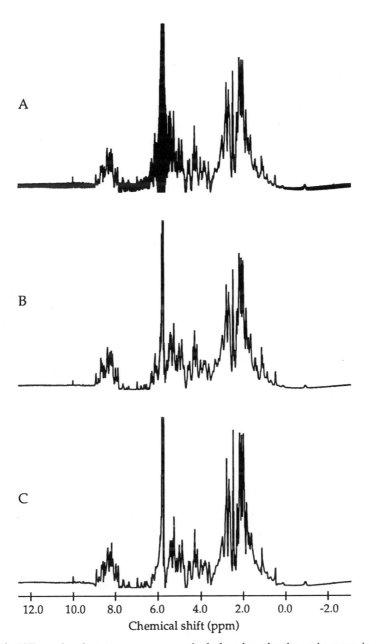

Fig. 6.4. IST can be slow to converge, particularly when the dynamic range is high. (**A**) shows the result of 240 iterations of IST. Truncation artifacts are still very evident. (**B**) is a MaxEnt reconstruction, shown for comparison. (**C**) is also the result of 240 iterations of IST, but this time using a variable threshold and explicit line search to minimize Q at each iteration.

240 steps, enough to converge within machine precision. Figure 6.4B shows the results of MaxEnt reconstruction for comparison. It turns out that the poor reconstruction in Figure 6.4A is due to lack of convergence and not the regularization functional, as can be seen by modifying the algorithm. Instead of iterating with a fixed value of the threshold, the initial threshold is taken to be a much larger value. The threshold is decreased until it reaches the final desired value. In addition, the step size at each iteration can be adjusted. Instead of moving by $-\nabla Q$, we can move by $-\beta \nabla Q$, where β is a scale factor chosen to make Q as small as possible; β can be found by a simple line search. The result of this algorithm is shown in Figure 6.4C. After using the same number of steps as Figure 6.4A, it is much closer to the MaxEnt reconstruction.

If IST and MaxEnt reconstruction give comparable, if slightly different re-sults, which should we use? Or should we use a completely different regular-ization functional, for that matter? It is difficult to give a satisfying answer to these questions. No one has yet come up with a regularization functional that corresponds to a spectroscopist's notion of "quality" (let alone one that also can be solved numerically): Until someone does, the best approach is to remain opportunistic: Use the method that seems to work best for the types of data you have. But don't forget that the burden of proof as to which method is most appropriate lies with the analyst; such considerations reside in the realm of error analysis, taken up in Chapter 7.

Having cast doubt on the proper choice of the regularization functional, we can go on to point out that there are algorithms for spectrum reconstruction that dispense with the regularization functional altogether. The Jansson-van Cittert algorithm [48] iteratively replaces the mock data with the actual data values, but does no regularization. Instead, the method relies on starting from a flat spectrum, in essence resulting in the spectrum that one obtains by fol-lowing ∇C from a perfectly flat spectrum to the fixed point.

Beyond the utility of the method itself, our introduction to IST emphasizes the points to be kept in mind when using *any* iterative spectrum reconstruction algorithm: the regularization functional (if there is one), convergence criteria, and efficiency. As with all nonlinear methods of spectrum analysis, *caveat emptor*. It has been said of one fixed-point method that it "requires almost no mathematical theory" [49]. True enough, perhaps, if one's sole desire is to use the method without understanding it. But the potential danger is clear: Beware the chef who uses a new spice without checking the taste!

6.2. SMOOTHING BY WAVELETS

Another image-based approach to data analysis, quite different from the types of methods we have considered so far, is based on wavelets (which we will define in a moment). We must admit that there have been very few applications of wavelets to NMR, but wavelets are a new and rapidly emerging—not to mention sexy—field. For example, one of the applications of wavelets is for image compression in the transmission of digital television signals. They have also been applied to other areas of signal processing, and we expect that it will not be long before those applications migrate to NMR. Here we give one simple example: spectrum smoothing.

Like the discrete Fourier transform, the *discrete wavelet transform* (DWT) is based on an expansion in terms of orthonormal basis vectors, called *wavelets*. Whereas the Fourier basis vectors are related to one another by frequency modulation, the wavelet basis vectors are related to one another by shifting (translation) and stretching (dilation).

The basic idea behind wavelets is to construct a basis that is sensitive to detail at a variety of scales. This is done recursively by constructing basis vectors that are sensitive to detail at one scale from basis vectors sensitive to detail on a shorter scale. As an example, consider the standard basis vectors for a signal consisting of eight samples

$$
\begin{aligned}
&(1, 0, 0, 0, 0, 0, 0, 0) \\
&(0, 1, 0, 0, 0, 0, 0, 0) \\
&(0, 0, 1, 0, 0, 0, 0, 0) \\
&(0, 0, 0, 1, 0, 0, 0, 0) \\
&(0, 0, 0, 0, 1, 0, 0, 0) \\
&(0, 0, 0, 0, 0, 1, 0, 0) \\
&(0, 0, 0, 0, 0, 0, 1, 0) \\
&(0, 0, 0, 0, 0, 0, 0, 1)
\end{aligned}
\tag{6.12}
$$

These basis vectors are related to one another by translation, but they are only sensitive to detail at one scale, namely that of a single point in the data vector.

Suppose the smallest scale we are interested in is detail spanning four consecutive data samples. We can construct suitable basis vectors by taking linear combinations of the standard basis vectors, such as:

$$
\begin{aligned}
&(d_1, d_2, d_3, d_4, 0, 0, 0, 0) \\
&(0, d_1, d_2, d_3, d_4, 0, 0, 0) \\
&(0, 0, d_1, d_2, d_3, d_4, 0, 0) \\
&(0, 0, 0, d_1, d_2, d_3, d_4, 0) \\
&(0, 0, 0, 0, d_1, d_2, d_3, d_4) \\
&(d_4, 0, 0, 0, 0, d_1, d_2, d_3) \\
&(d_3, d_4, 0, 0, 0, 0, d_1, d_2) \\
&(d_2, d_3, d_4, 0, 0, 0, 0, d_1)
\end{aligned}
\tag{6.13}
$$

These basis vectors are all translations of one another, too (note the circular translation). We can ensure that the vectors are orthogonal to one another by suitable choice of the coefficients d_1, d_2, d_3, and d_4. The problem with this basis is that it is only sensitive to detail at the scale of four adjacent points. We will have to give up some of the vectors if we want to probe detail at more than one scale.

Let's give up half of the vectors, keeping only translations by even amounts. This makes a certain amount of sense, since we then at least have basis vectors sensitive to features four points wide along the entire length of the data vector; all we are giving up is some precision about the location of the features. So our first four basis vectors are

$$(d_1, d_2, d_3, d_4, 0, 0, 0, 0)$$
$$(0, 0, d_1, d_2, d_3, d_4, 0, 0)$$
$$(0, 0, 0, 0, d_1, d_2, d_3, d_4)$$
$$(d_3, d_4, 0, 0, 0, 0, d_1, d_2)$$

(6.14)

and we will use the remaining four to probe longer scales. For the first four basis vectors to be orthogonal, the coefficients must satisfy the equation $d_1 d_3 + d_2 d_4 = 0$. It turns out that there are four other vectors, closely related to these four, that are orthogonal to all of these and to each other, as well as being translations by two. They arise from reversing the coefficients and alternating signs:

$$(-d_4, d_3, -d_2, d_1, 0, 0, 0, 0)$$
$$(0, 0, -d_4, d_3, -d_2, d_1, 0, 0)$$
$$(0, 0, 0, 0, -d_4, d_3, -d_2, d_1)$$
$$(-d_2, d_1, 0, 0, 0, 0, -d_4, d_3)$$

(6.15)

These basis vectors form a so-called *quadrature mirror* of the ones in display (6.14). It's easy to see that these vectors are orthogonal to one another if the first set of vectors is orthogonal (the necessary condition is again $d_1 d_3 + d_2 d_4 = 0$), and that the two sets are mutually orthogonal. While we don't want to keep both sets in our final basis, the second set is useful for constructing a linear combination that is sensitive to detail at a different scale. By constructing the new basis vectors from linear combinations of vectors that are orthogonal to those we *are* keeping, we ensure that the new basis vectors are also orthogonal to the ones we keep.

What linear combination should we use? Here's where recursion comes in: We can apply the same procedure we used to go from scale one to scale four. Let W_1, W_2, W_3, W_4 be the four vectors in display (6.15); then two of the vectors with scale eight are given by:

$$d_1\mathbf{W}_1 + d_2\mathbf{W}_2 + d_3\mathbf{W}_3 + d_4\mathbf{W}_4$$
$$d_3\mathbf{W}_1 + d_4\mathbf{W}_2 + d_1\mathbf{W}_3 + d_2\mathbf{W}_4 \tag{6.16}$$

(again, notice the circular translation). The remaining two basis vectors are the quadrature mirror of these:

$$-d_4\mathbf{W}_1 + d_3\mathbf{W}_2 - d_2\mathbf{W}_3 + d_1\mathbf{W}_4$$
$$-d_2\mathbf{W}_1 + d_1\mathbf{W}_2 - d_4\mathbf{W}_3 + d_3\mathbf{W}_4 \tag{6.17}$$

At this point we have run out of vectors; there are only two quadrature mirror vectors remaining, which isn't enough to continue the recursion. This same idea could be applied to longer vectors, so long as the length is a power of two.

Although we know the basic strategy, we still haven't specified the values of the coefficients. To do so we need to introduce some additional conditions. Daubechies imposed the requirements that the wavelets [the basis vectors in (6.14)] be orthogonal to (1, 1, 1, 1, 1, 1, 1, 1) and (1, 2, 3, 4, 5, 6, 7, 8) [50]. This means that a constant vector or one that is linear will have no component along the basis vectors of scale four, and it is equivalent to requiring that the first two moments of the wavelet vectors vanish. An additional normalization condition $(d_1^2 + d_2^2 + d_3^2 + d_4^2 = 1)$ brings us to four equations in four unknowns; the solution is

$$d_1 = \frac{1}{4\sqrt{2}}(\sqrt{3} - 1) \qquad d_2 = \frac{1}{4\sqrt{2}}(3 - \sqrt{3})$$
$$d_3 = \frac{1}{4\sqrt{2}}(-3 - \sqrt{3}) \qquad d_4 = \frac{1}{4\sqrt{2}}(\sqrt{3} + 1) \tag{6.18}$$

This is the Daubechies D4 wavelet basis, and it is just one of a family of bases. Other members of the family have more nonzero components at the lowest level and additional constraints imposed on the coefficients. For example, the D6 wavelet basis has six nonzero components at the lowest level, and the first three moments vanish.

Some of the D4 basis vectors (of length 256) are shown in Figure 6.5. Figure 6.5A shows a basis vector from the lowest level of detail, with four nonzero components. The basis vectors in Figure 6.5B and C are from successively higher scales, and it is clear that the basis vectors in Figure 6.5A–C are related to one another by dilation and translation. Figure 6.5D shows one of the two final basis vectors (sometimes called the mother function). Since these vectors were left over at the end of the rec sion scheme, they are not dilated or translated versions of the wavelet.

There are wavelets other than the ones Daubechies invented. In fact, there is an infinite number of ways to construct wavelets, which affords a great deal

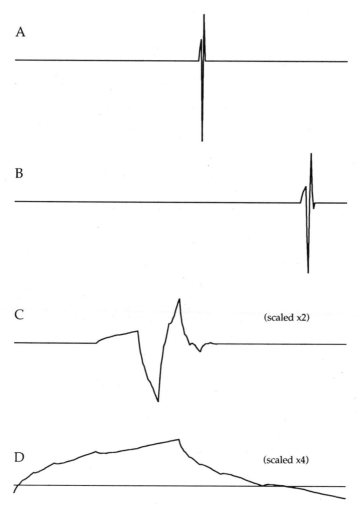

Fig. 6.5. These representative members of the Daubechies D4 wavelet basis (for 256-point vectors) span scales of length (**A**) 4, (**B**) 8, (**C**) 64, and (**D**) 256. Notice that the vectors in (**A–C**) are related by translation and dilation. In contrast to the Fourier basis vectors (sines and cosines), which are localized in frequency but not in time, the wavelet basis vectors are localized in both time and frequency.

of flexibility, in that wavelets can be constructed to match characteristics of particular classes of signals. For example, Figure 6.6 shows basis vectors of the "nearly-symmetric" wavelet S4 (which has eight coefficients) [51]. In principle, a more symmetric wavelet is better suited to characterizing the symmetric peaks that occur in NMR spectra. Unfortunately, there is no perfectly symmetric wavelet.

As we promised earlier, we now show how wavelets can be used for smoothing (an application originally suggested to us by David Donoho). The procedure consists of applying a discrete wavelet transform to the spectrum, performing soft thresholding in the wavelet domain to eliminate noise, and inverting the wavelet transform. The various stages are illustrated in Figure 6.7. Notice that the noise is reduced without broadening the peaks or reducing their intensity,

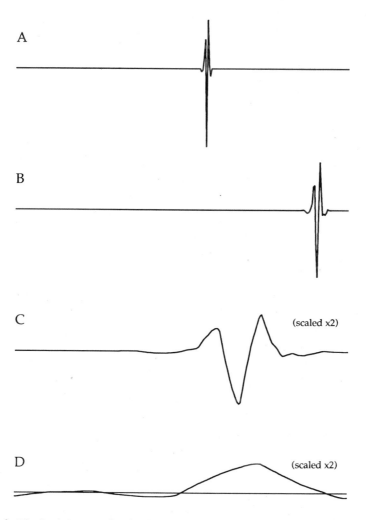

Fig. 6.6. The basis vectors for the S4 "nearly symmetric" wavelet basis resemble the D4 wavelet vectors in Figure 6.5, but are smoother and more symmetric. In spite of the resemblance, the S4 basis vectors have eight nonzero components at the lowest level. (**A–D**) contain vectors corresponding to the ones in Figure 6.5.

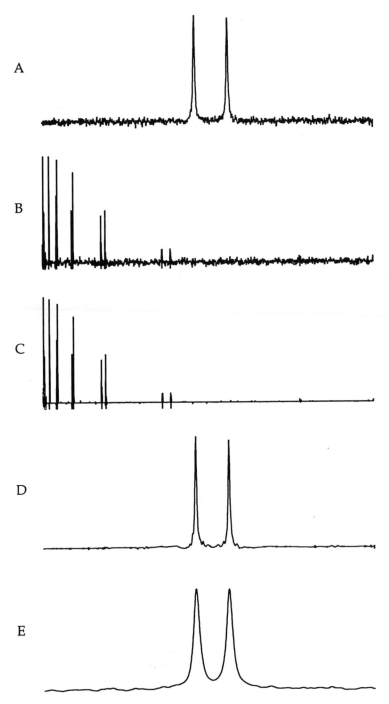

Fig. 6.7. The discrete wavelet transform can be used to reduce noise in the spectrum without broadening peaks. (**A**) shows a synthetic spectrum containing noise. In (**B**) the

in contrast to noise reduction by exponential apodization. (Incidentally, the small number of nonzero wavelet coefficients needed to reconstruct the spectrum hints at how wavelets can be used for data compression.) The smooth spectrum is not ideal; it contains some residual features that reflect the structure of the wavelet basis vectors. These features may or may not be objectionable, depending on the application. It would be an understatement to say that the use of wavelets in NMR is not widespread, but we anticipate further developments as more NMR spectroscopists become aware of the unusual properties of wavelets.

6.3. BAYESIAN TECHNIQUES

Bayesian analysis is a statistical technique for testing the degree to which an hypothesis is confirmed by experimental data. Before 1990 it had not been widely applied in NMR, but that has begun to change, thanks to the efforts of such people as Bretthorst [52] and Sibisi [53]. Although the details can be quite complicated, the basic idea behind Bayesian analysis is very simple: It is nothing more than probability theory. We will start by reviewing some common notation.

In probability theory, letters like A and B stand for particular propositions. Since we are concerned with NMR, we will use such symbols as $\mathbf{d}$ to represent the statement that the measured FID is equal to $\mathbf{d}$, or w for the statement that a sinusoid has frequency equal to w. These statements will have various probabilities; $P(A)$ stands for the probability of the proposition A. You can regard this probability as a measure of how strongly you believe that A is true, or as an indication of the fraction of the outcomes in which A would be true if the experiment were repeated many times.

Similarly, $P(A, B)$ is the *joint* probability that A and B are both true, and $P(A|B)$ is the *conditional* probability of A given that B is true. (This can be thought of as how strongly you would believe that A were true if you knew that B was indeed true. Alternatively, if the experiment were repeated many times, this probability would represent the fraction of the outcomes in which A and B were both true, out of all those in which B was true.)

Incidentally, there are some deep philosophical issues connected with the notion of probability in science. The "degree of belief" and the "frequency of occurrence" interpretations are not really the same. For example, if you are 90% certain that object A is heavier than the object B, that does *not* mean you

---

spectrum has been transformed to the wavelet domain (using the S4 wavelet). (**C**) shows the result of soft thresholding in which the smallest 90% of the wavelet coefficients have been set to zero. (**D**) shows the spectrum following inverse wavelet transformation of the thresholded data. For comparison, the result of conventional apodization using the matched filter is plotted in (**E**).

expect that nine out of ten measurements will give a larger weight for A. We will blithely skip over these profound matters, however; such distinctions are too subtle to affect our discussion.

There are two fundamental laws of probability essential to the analysis. The first says the probability that any one of a class of disjoint (i.e., mutually exclusive) propositions will be true is just the sum of the individual probabilities (or for a continuous class of propositions, the integral). For example, suppose $P(w, \phi)$ is the probability that an FID contains a signal with frequency w and phase ϕ. Differing phases do indeed represent disjoint propositions (a signal cannot have *two* phases), so this law tells us that the probability of the FID containing a signal with frequency w, regardless of the phase, is given by

$$P(w) = \int_{-\pi}^{\pi} P(w, \phi) \, d\phi \qquad (6.19)$$

The second law concerns conditional probabilities. It says the joint probability of two propositions both being true is equal to the product of the individual probability that the first proposition is true, times the conditional probability of the second proposition given the first. Symbolically,

$$P(A, B) = P(A)P(B|A) \qquad (6.20)$$

Of course, this makes perfectly good sense if you think of the probabilities as meaning the fraction of the time that the propositions are true.

One last definition is worth mentioning. Two propositions are *independent* if their joint probability is equal to the product of their individual probabilities. That is, A and B are independent if

$$P(A, B) = P(A)P(B) \qquad (6.21)$$

You can easily see that this is the same as saying that the individual probabilities are the same as the conditional probabilities:

$$P(A) = P(A|B) \quad \text{or} \quad P(B) = P(B|A) \qquad (6.22)$$

Bayesian analysis is based on *Bayes' rule*, which is just an application of Eq. (6.20). Let D stand for the value of the data, and let $\mathfrak{M}$ stand for a particular model. Then

$$P(D)P(\mathfrak{M}|D) = P(D, \mathfrak{M}) = P(\mathfrak{M})P(D|\mathfrak{M}) \qquad (6.23)$$

and so

$$P(\mathfrak{M}|D) = \frac{P(D|\mathfrak{M})P(\mathfrak{M})}{P(D)} \qquad (6.24)$$

Equation (6.24) is Bayes' rule.* It shows us how to calculate the probability of our model being true, given the experimentally measured data. In practice, we most often have a whole collection of possible models in mind; the rule then lets us compare their probabilities so we can select the model most likely to be true, in view of the data. For this purpose, it is convenient to drop the $P(D)$ term (which does not depend on the model, after all) and write the rule as

$$P(\mathfrak{M}|D) \propto P(D|\mathfrak{M})P(\mathfrak{M}) \qquad (6.25)$$

The quantity $P(\mathfrak{M}|D)$ is called the *posterior* probability of the model $\mathfrak{M}$ (its probability once the data are known), $P(D|\mathfrak{M})$ is called the *likelihood* of the data (how likely it is that we would have obtained the actual experimental values if the model were true), and $P(\mathfrak{M})$ is called the *prior* probability of $\mathfrak{M}$ (how strongly we believe that the model could be true based just on our general knowledge, before knowing the results of the experiment). Among other things, Eq. (6.25) says that if the data are explained equally well by either of two models (that is, if the likelihoods are the same), but one model is inherently less probable than the other, we should give greater credence to the more probable model. "Extraordinary conclusions require extraordinary evidence" is a philosophy that certain scientists of recent painful memory would have done well to bear in mind (remember polywater and cold fusion?).

Let's see how to apply all this to NMR. We will only discuss one-dimensional NMR, for simplicity, but the concepts can easily be applied to multidimensional experiments [54]. Our model for an FID is a sum of L exponentially decaying sinusoids, plus Gaussian random noise:

$$d_k = \sum_{j=1}^{L} [(A_j e^{i\phi_j}) e^{-k\Delta t/\tau_j} e^{2\pi i k \Delta t w_j}] + \varepsilon_k, \qquad k = 0, \ldots, M-1 \quad (6.26)$$

where A_j is the amplitude, ϕ_j the phase, τ_j the decay time, and w_j the frequency of the jth sinusoid, Δt is the sampling interval, and ε_k is Gaussian random noise with zero mean and standard deviation σ. Again, for simplicity we will assume that the standard deviation of the noise is the same for each point in the FID, although this is not necessary. We are also assuming that the data points have been linearly sampled (i.e., the interval between samples is a uniform Δt), but this need not be true either.

The parameters of the model are $\mathbf{A}$, ϕ, τ, $\mathbf{w}$, and σ. The noise values ε are not determined solely by the model; rather they are equal to the difference between the sum of decaying sinusoids and the actual FID. Given this model,

---

*Many people like to include in their probability formulas the factor I, which stands for all the prior information known before any experiments are performed. Doing so adds nothing to our discussion, and we will omit the I's.

what is the likelihood of observing a particular FID? Since the parameters of the model specify the systematic component of the signal exactly, only the ε values are free to vary. Consequently, the likelihood of a particular FID is nothing more than the likelihood of the corresponding ε. We are assuming that each ε_k has a normal distribution, so that

$$P(\varepsilon_k|\sigma) = \frac{1}{2\pi\sigma^2} e^{-|\varepsilon_k|^2/2\sigma^2} \tag{6.27}$$

(The coefficient is $1/2\pi\sigma^2$ rather than $1/\sqrt{2\pi\sigma^2}$ because ε_k includes both a real and an imaginary component.) Since the noise values are assumed to be independent, the joint probability of all the ε_k's is just the product of their individual probabilities. Thus, the likelihood of the data $\mathbf{d}$ is given by

$$P(\mathbf{d}|\mathbf{A}, \phi, \tau, \mathbf{w}, \sigma, L) = \prod_{k=0}^{M-1} P(\varepsilon_k|\mathbf{A}, \phi, \tau, \mathbf{w}, \sigma, L)$$

$$= \frac{1}{(2\pi\sigma^2)^M} \exp\left(\sum_{k=0}^{M-1} -|\varepsilon_k|^2/2\sigma^2\right) \tag{6.28}$$

where ε_k is determined by Eq. (6.26).

The phase ϕ_j appears in an exponent, but by introducing the definition

$$\alpha_j = A_j e^{i\phi_j} \tag{6.29}$$

we can reduce the amplitude A_j and phase ϕ_j to a single complex variable α_j. Equation (6.26) is linear in the α's, so it is possible to solve for them explicitly and eliminate them. Let $\mathbf{U}$ be the $(M \times L)$ model matrix given by

$$U_{k,j} = e^{-k\Delta t/\tau_j} e^{2\pi i k \Delta t w_j} \tag{6.30}$$

Now our model can be expressed by the single matrix equation

$$\mathbf{d} = \mathbf{U}\boldsymbol{\alpha} + \boldsymbol{\varepsilon} \tag{6.31}$$

and the likelihood is equal to

$$P(\mathbf{d}|\boldsymbol{\alpha}, \tau, \mathbf{w}, \sigma, L)$$

$$= \frac{1}{(2\pi\sigma^2)^M} \exp\left(\sum_{k=0}^{M-1} -|d_k - (\mathbf{U}\boldsymbol{\alpha})_k|^2/2\sigma^2\right) \tag{6.32}$$

Using the laws of probability, we can eliminate the α's by computing

$$P(\mathbf{d}|\tau, \mathbf{w}, \sigma, L) = \int P(\mathbf{d}, \boldsymbol{\alpha}|\tau, \mathbf{w}, \sigma, L) \, d\boldsymbol{\alpha}$$

$$= \int P(\mathbf{d}|\boldsymbol{\alpha}, \tau, \mathbf{w}, \sigma, L) P(\boldsymbol{\alpha}|\tau, \mathbf{w}, \sigma, L) \, d\boldsymbol{\alpha} \quad (6.33)$$

Unfortunately, we do not have a prior probability for $\boldsymbol{\alpha}$ [the last factor in Eq. (6.33)]. Lacking any other information, we would like to be as noncommittal as possible about the values of the amplitudes and phases. For now we will simply take $P(\boldsymbol{\alpha}|\tau, \mathbf{w}, \sigma, L)$ to be equal to one, corresponding to a flat distribution. (In fact, this is not a true probability distribution at all, since its integral over all values of $\boldsymbol{\alpha}$ is infinity. It is a so-called *improper* prior, and later on we will have to modify it.)

To evaluate the integral in Eq. (6.33), start by setting Q to be the sum of the squared errors:

$$Q = \sum_{k=0}^{M-1} |d_k - (\mathbf{U}\boldsymbol{\alpha})_k|^2 = (\mathbf{d} - \mathbf{U}\boldsymbol{\alpha})^{\dagger}(\mathbf{d} - \mathbf{U}\boldsymbol{\alpha})$$

$$= \mathbf{d}^{\dagger}\mathbf{d} - (\mathbf{U}\boldsymbol{\alpha})^{\dagger}\mathbf{d} - \mathbf{d}^{\dagger}\mathbf{U}\boldsymbol{\alpha} + (\mathbf{U}\boldsymbol{\alpha})^{\dagger}\mathbf{U}\boldsymbol{\alpha} \quad (6.34)$$

$$= \mathbf{d}^{\dagger}\mathbf{d} - 2 \cdot \text{real } (\mathbf{d}^{\dagger}\mathbf{U}\boldsymbol{\alpha}) + \boldsymbol{\alpha}^{\dagger}\mathbf{U}^{\dagger}\mathbf{U}\boldsymbol{\alpha}$$

The "interaction matrix" $\mathbf{U}^{\dagger}\mathbf{U}$ can be diagonalized; that is, we can write

$$\mathbf{U}^{\dagger}\mathbf{U} = \mathbf{E}\boldsymbol{\Lambda}\mathbf{E}^{\dagger} \quad (6.35)$$

where $\mathbf{E}$ is the matrix of eigenvectors of $\mathbf{U}^{\dagger}\mathbf{U}$ and $\boldsymbol{\Lambda}$ is a diagonal matrix containing the eigenvalues $\lambda_1, \ldots, \lambda_L$. Set $\mathbf{b} = \mathbf{E}^{\dagger}\boldsymbol{\alpha}$, so that in turn, $\boldsymbol{\alpha} = \mathbf{E}\mathbf{b}$. Then we have

$$Q = \mathbf{d}^{\dagger}\mathbf{d} - 2 \cdot \text{real}(\mathbf{d}^{\dagger}\mathbf{U}\mathbf{E}\mathbf{b}) + \mathbf{b}^{\dagger}\boldsymbol{\Lambda}\mathbf{b}$$

$$= \sum_{k=0}^{M-1} |d_k|^2 + \sum_{j=1}^{L} \{-2 \cdot \text{real}[(\mathbf{d}^{\dagger}\mathbf{U}\mathbf{E})_j b_j] + \lambda_j |b_j|^2\} \quad (6.36)$$

Let $\mathbf{h}$ be the vector defined by

$$h_j = (\mathbf{d}^{\dagger}\mathbf{U}\mathbf{E})_j^* / \lambda_j \quad (6.37)$$

then Q reduces to

$$Q = \sum_{k=0}^{M-1} |d_k|^2 + \sum_{j=1}^{L} \{-2\lambda_j h_j^r b_j^r - 2\lambda_j h_j^i b_j^i + \lambda_j (b_j^r)^2 + \lambda_j (b_j^i)^2\} \quad (6.38)$$

where the superscripts r and i refer to the real and imaginary components. The Jacobian $|d\alpha/d\mathbf{b}|$ is equal to one because the matrix $\mathbf{E}$ is unitary. So changing the variable of integration from α to $\mathbf{b}$, the integral in Eq. (6.33) yields

$$
\begin{aligned}
P(\mathbf{d}|\tau, \mathbf{w}, \sigma, L) &= \frac{1}{(2\pi\sigma^2)^M} \int \exp(-Q/2\sigma^2)\, d\mathbf{b} \\
&= (2\pi\sigma^2)^{L-M}(\lambda_1 \cdots \lambda_L)^{-1} \exp\left(\frac{-M\overline{d^2} + L\overline{h^2}}{2\sigma^2}\right)
\end{aligned}
\tag{6.39}
$$

where

$$
\overline{d^2} = \frac{1}{M}\sum_{j=0}^{M-1}|d_j|^2 \quad \text{and} \quad \overline{h^2} = \frac{1}{L}\sum_{k=1}^{L}|h_k|^2
\tag{6.40}
$$

are the average squared absolute values of $\mathbf{d}$ and $\mathbf{h}$.

If the noise level σ and the number of sinusoids L are known beforehand, Eqs. (6.39) and (6.40) show how to calculate the probabilities of various possible combinations of decay times and frequencies. A straightforward fitting procedure can be used to find the values that yield the highest probability. Once these are have been found, the most likely values of the amplitudes and phases can be computed by maximizing Q; the result is that $\mathbf{b} = \mathbf{h}$ and so $\alpha = \mathbf{Eh} = \Lambda^{-1}\mathbf{U}^\dagger\mathbf{d}$.

If σ and L are *not* known beforehand, it is still possible (although more difficult) to carry out the analysis. A prior probability distribution for σ must be chosen, and the improper prior we used for the α's must be replaced with a proper one. Bretthorst performed the calculation in [52], using maximum entropy probability distributions—the most noncommittal ones available. He showed that the likelihood is given by the formula

$$
P(\mathbf{d}|\tau, \mathbf{w}, L) \propto \Gamma(L)\,\Gamma(M-L)\left(\frac{L\overline{h^2}}{2}\right)^{-L}\left(\frac{M\overline{d^2} - L\overline{h^2}}{2}\right)^{L-M}
\tag{6.41}
$$

where $\Gamma(x)$ is the gamma function of x. The dependence on the decay times and frequencies can be eliminated by numerical integration (after weighting the likelihood by an appropriate prior); however, it may be easier to cheat and simply use the values giving the highest likelihood. For the w's, at least, this is a reasonable approximation, because NMR data usually determine the resonance frequencies fairly precisely. The τ's, on the other hand, often cannot be determined so well. At any rate, once this has been done, L can be chosen simply by trying out different numbers to see which is most likely. If you have some prior expectation of how many sinusoids there should be, that can be taken into account using Bayes' rule:

$$
P(L|\mathbf{d}) \propto P(\mathbf{d}|L)P(L)
\tag{6.42}
$$

TABLE 6.1. Results from Bayesian Spectrum Analysis of the FID Used to Generate Figure 6.8A.

Frequency (Hz)	Amplitude	Phase (degrees)	Line Width (Hz)
−597	10.9	62	>5.0
−604	29.8	66	4.5
−612	26.6	65	3.0
−620	9.2	82	3.4
−788	6.0	27	>5.0
−795	2.2	−19	<0.5
−801	84.0	71	1.7
−812	6.1	97	>5.0
−1054	32.4	67	2.7
−1165	30.1	75	2.7
−1173	64.1	73	3.5
−1180	36.5	57	5.0

Bayesian analysis is similar to the methods based on linear prediction in that it directly provides a parametric description of the spectrum. In addition, since it gives probabilities, it also yields error estimates for the parameters. This allows one to make such statements as, "With 95% probability, there exists a signal with frequency in the range 1531 ± 4 Hz." Using this information, one can present the spectrum in a rather unusual fashion, by plotting for each frequency the probability that there is a signal at that frequency, rather than the intensity of the signal.

To emphasize the parametric nature of Bayesian analysis, an example of the results of the method (using the program CHIFIT written by Roger Chylla) is shown in Table 6.1. Conventional spectra in the form of plots of intensity as a function of frequency are generated by plugging these parameter values into the systematic part of the model, a sum of decaying sinusoids (Fig. 6.8). Note that parameters for some of the peaks listed in Table 6.1 have markedly different phases or line widths from the others. If the phase or line width of the real peaks is known *a priori*, peaks that have sufficiently different parameters are often attributed to noise, and left out of the reconstructed spectrum.

6.4. MULTIPLE-TAPER SPECTRUM ANALYSIS

To conclude this chapter, we will briefly mention a little-known generalization of the matched filter. We described the matched filter in Chapter 3 and mentioned that it is the optimal window function for enhancing the S/N. Multiple taper spectral analysis takes advantage of the fact that it is possible to construct a family of *nearly* optimal window functions that are orthogonal to each other and to the matched filter. Because of this orthogonality, each time series that results from using one of these nearly optimal window functions can be viewed

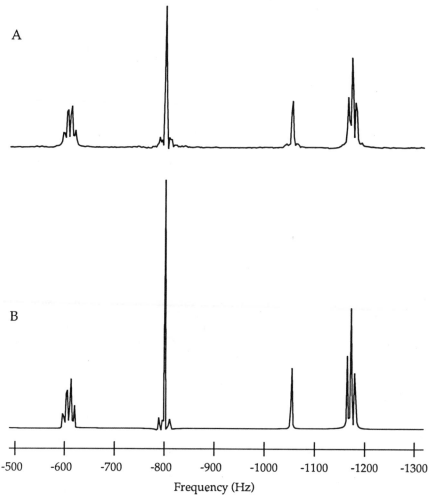

Fig. 6.8. Bayesian spectrum analysis yields a list of parameter values: frequencies, amplitudes, phases, and line widths. These values can be converted into a conventional spectrum by adding together a collection of Lorentzian peaks constructed from the parameters. (**A**) contains a DFT spectrum, and (**B**) contains a spectrum synthesized using the parameters in Table 6.1. (A phase correction was applied to the spectra, so the peaks do not appear with the phases listed in the table.) Features not identified by the Bayesian analysis (such as ones comparable in intensity to the noise) do not appear in the reconstructed spectrum.

as an independent measurement of the data. This means that a statistical F-test, for example, could be applied to these independent data sets to estimate the probability that a sinusoid with a particular frequency is present. It has been shown that this approach is a more sensitive means of detecting decaying sinusoids in the presence of noise than using a single window function [55].

TO READ FURTHER

Strang has written an informal introduction to wavelets [56], and *Numerical Recipes in Fortran* [24] provides algorithms for computing the discrete wavelet transform. The books by Papoulis [57] and Oppenheim and Schafer [58] cover a broad range of topics in signal processing and spectrum analysis.

7

VISUALIZATION, QUANTIFICATION, AND ERROR ANALYSIS

We conclude the book with a brief look at three aspects of NMR data processing that are often taken for granted. The treatments in this chapter are not rigorous; if the first chapter could be viewed as an appetizer, then this is a light dessert. Nevertheless, the subjects are worthy of consideration by anyone who performs NMR data processing.

We've expounded at considerable length about ways to estimate the spectrum based on measurement of the FID, but we've said relatively little about how to present the results. Visualization is the most effective way to discern *qualitative* features of spectra. We have used one-dimensional plots and two-dimensional contour plots up to now without commenting on details of plotting data. Plotting one-dimensional data is straightforward, but visualizing multi-dimensional data is more challenging and worthy of some discussion. Some of the facets of visualization we'll point out are obvious, others subtle, but all are useful and a few may even be profound.

Quantitative use of NMR spectra involves numerically estimating characteristics of spectral components, such as frequency and amplitude. We'll describe some simple methods for quantifying features of spectra obtained from nonparametric techniques (DFT, MaxEnt reconstruction, etc.). Of course, parametric techniques (LP, Bayesian analysis) directly provide numerical values for spectral features.

Equally important are methods for assessing accuracy and precision of quantitative results. Quantitation without error analysis is a bit like driving a car without wearing a seat belt; it may get you where you want to go, but the possibility of disaster lurks at every intersection. It is the responsibility of every analyst to have at least some idea of the errors in their results.

7.1. VISUALIZATION

The human visual cortex is a highly complex system, optimized through millions of years of evolution for the efficient detection of subtle characteristics of the visual field, ranging from shape and symmetry to orientation, motion, and color. It makes sense to try to exploit this biological machinery for the analysis of NMR data. The clever analyst will design visualization tools with characteristics of the visual system in mind.

The dimensionality of an NMR spectrum is determined by the experiment; one-, two-, and three-dimensional spectra are the most common. Four- and higher-dimensional spectra are appearing in some contexts. The "dimensionality" of the human visual system is not arbitrary, however, and one of the most important aspects of visualization is mapping dimensions of the data to dimensions of the visual system. The dimensions of multidimensional NMR spectra are all equivalent in a certain sense, since they all represent frequencies, but the dimensions of the visual field are not. Among the "dimensions" of the visual field are vertical and horizontal position, depth, and color. Two of the dimensions, horizontal and vertical position, are essentially equivalent, but depth perception is distinctly different, since it comes from perspective, stereopsis, and other visual cues. So is color perception, which derives not only from the wavelengths of light emanating from parts of the visual field, but also from the relationships between them.

7.1.1. One-dimensional Spectra

A one-dimensional spectrum has one independent and one dependent degree of freedom, so an ordinary line graph suffices to represent the spectrum. Most often, horizontal position is used to represent the independent degree of freedom (the frequency), and vertical position is used to represent the dependent degree of freedom (the intensity of the response). Sometimes a spectrum is represented by distinct, unconnected points rather than a line, in order to emphasize its discrete nature.

7.1.2. Two-dimensional Spectra

Two-dimensional spectra have two independent degrees of freedom and one dependent degree of freedom. One of the most useful ways to present two-dimensional data is the *contour plot*. The frequencies are represented by horizontal and vertical position, and the intensity as a function of the two frequencies is represented by lines connecting regions of constant intensity. This is exactly the same as a topographic contour map, in which the contours correspond to constant elevation. The spacing between contours gives an indication of the steepness of the underlying features: Closely spaced contours indicate a rapid change in intensity, widely spaced contours indicate a slow change in intensity.

Since the spectrum is discrete, in order to obtain continuous contours the intensity must be interpolated between points that straddle a given contour level. The algorithm for calculating contour lines is rather simple. Consider four neighboring points of the spectrum (i, j), $(i + 1, j)$, $(i + 1, j + 1)$, $(i, j + 1)$ (Fig. 7.1A), and a contour level C chosen so that the spectrum value $f_{i,j}$ is larger than C and $f_{i,j+1}$, $f_{i+1,j}$, and $f_{i+1,j+1}$ are smaller than C. Then only two sides of the box connecting the four points of the spectrum cross the contour level. Linear interpolation is used to determine the intersection points, E and F. The point E, for example, is given by

$$\left(i, j + \frac{C - f_{i,j}}{f_{i,j+1} - f_{i,j}}\right) \tag{7.1}$$

Point F is determined in a similar manner. The line segment EF is drawn to represent the contour.

It's easy to see that for an arbitrary contour level, the number of sides of the box that intersect the level will be zero, two, or four (although due care must be exercised when a contour level exactly matches the data value at one of the corner points). If there are no crossings, nothing is drawn. If all four sides intersect the contour level, there are six possible line segments connecting the intersection points. Contour lines should never cross each other, so there are two ways of connecting the intersection points when there are four crossings (Fig. 7.1B and C). The choice of which to use is arbitrary, since the data are insufficient to determine the correct possibility, but the choice should be made consistently.

To complete the contour plot, this procedure is repeated for each quadruple of adjacent points in the two-dimensional spectrum. The line segments for adjacent boxes will match up, and the contours either will be closed curves or will terminate at the edges of the plot. It's possible to add the additional wrinkle of connecting the intersection points with smooth curves (splines), rather than line segments, which produces nicer-looking contour plots when the grid points are widely spaced. If the spacing is fairly small, you probably won't notice the difference, and it's not worth the additional computational effort.

For cathode ray tube (CRT) displays, laser printers, or other display devices that exact no penalty for drawing the line segments in discontinuous order, a fast algorithm treats quadruples in the order that they are laid out in computer memory. A really efficient algorithm will take advantage of the fact that each pair of interior points marks the edge of two adjacent boxes, and will only perform the calculation of the intersection points once.

For mechanical display devices, such as pen or ink-jet plotters, there *is* a penalty for drawing line segments in discontinuous order. The penalty arises from the time it takes to physically position the pen or print head, and from excessive up-and-down pen movements, which can exacerbate wear and tear.

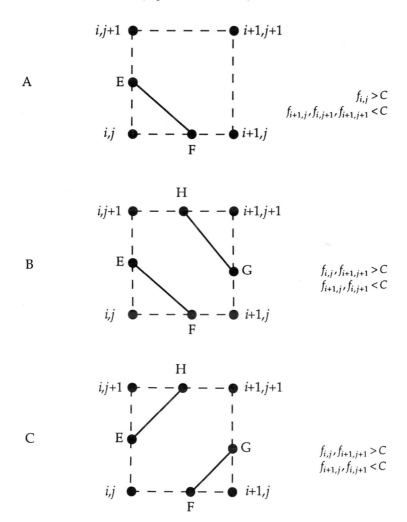

Fig. 7.1. Contour plots are constructed by plotting line segments between pairs of intersection points: places where a linear interpolation between adjacent points in the spectrum matches a contour level. In (**A**), only the edges between $(i, j) - (i, j + 1)$ and $(i, j) - (i + 1, j)$ cross the contour level, so there are only two intersection points, E and F. In (**B**) and (**C**), all four edges cross the contour level, so there are two possible ways of connecting the intersection points E, F, G, and H without allowing the lines to cross.

Up-and-down movements and the total distance the pen travels can be minimized by sorting the line segments so that they are drawn in continuous order. Although the time that it takes to compute line segments by tracing them until they close on themselves or they reach an edge is considerably longer than the time needed to compute them in sequential order, the computational cost is usually minor compared to the time saved in the physical plotting process.

A legacy of the widespread use of pen plotters is that some programs use a contour-following algorithm regardless of the display device. The next time you generate a contour plot on a CRT display, note the order in which the line segments are drawn. If they start at one edge of the plot and move across, a "fast" algorithm is being used. If, instead, entire closed contours are drawn in succession, a "pen plotter" algorithm is being used: admirable, but dumb! (Some programs buffer the output, waiting to display the vectors until they have all been computed. They all appear at once, and you can't tell which algorithm is being used.)

An alternative two-dimensional representation can be constructed by using perspective and other depth cues. This method of visualization goes by such names as stacked plot, hidden surface plot, or visible surface plot. Instead of computing interpolated two-dimensional line segments, the three-dimensional line segments that connect adjacent points in the spectrum are computed and then drawn in two-dimensional perspective. Sometimes only the lines in the X direction are drawn; sometimes lines in both the X and Y directions are drawn. The resulting view rather effectively conveys depth. An additional cue to the three-dimensional nature of the spectrum can be provided by making the line segments less intense, or using thinner line segments, as the distance from the viewer increases; this technique is called *depth cueing*. Another possibility is to treat the spectrum as being made up of opaque "tiles," whose edges are the three-dimensional line segments, with only exposed surfaces visible. A relatively straightforward algorithm for hidden surface removal works by sorting the edges by distance from the viewer. This is called a z-buffer algorithm, and special-purpose hardware for z-buffering (in nearly real time) is now widely available. To paraphrase Fermat, we don't have room to do justice to the fundamental mathematics of computer graphics, which in any event is covered in great detail elsewhere (a good place to start is the book by Foley et al. [59]).

A virtue of perspective representations of two-dimensional data is that they can readily portray, albeit qualitatively, the relative magnitude of peaks, noise, and artifacts (Fig. 7.2). They are not very useful for quantitative comparisons of magnitude or position, however, since their appearance depends on the orientation of the spectrum relative to the viewer.

7.1.3. Color as a Third Dimension

Something closer to a truly three-dimensional display of the two independent and one dependent dimensions of a two-dimensional spectrum can be obtained by using color to represent the third dimension. The frequency dimensions are represented by horizontal and vertical position, and the intensity of the spectrum is represented by the color of a spot, called a picture element (pixel for short), placed at the appropriate position. The correspondence between the data values and color is often illustrated using a "color bar," as shown at the right in Figure 7.3. One advantage of pseudocolor representations (as they are called) over contour plots is that it is easier to quickly discern the relative intensities

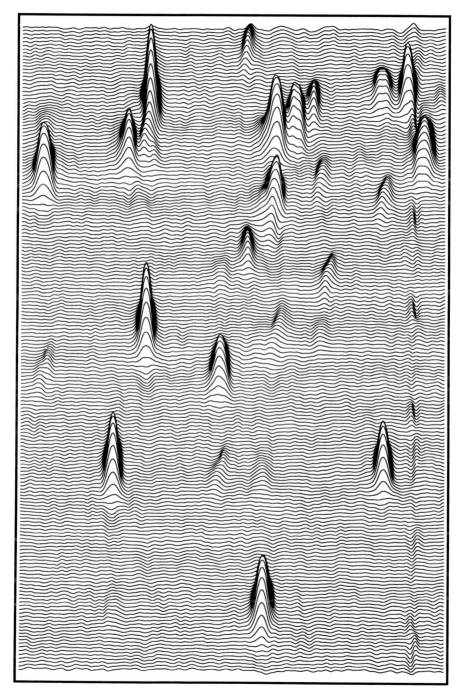

Fig. 7.2. In a ''hidden surface'' plot of two-dimensional data, the spectrum is treated as an opaque surface. Portions of the data lying behind other features are not drawn.

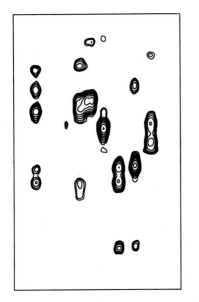 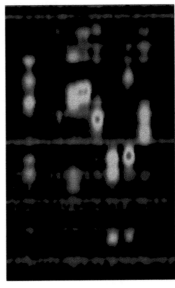 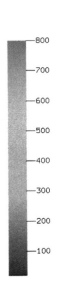

Fig. 7.3. Two-dimensional spectra can be plotted using color to represent intensity. Here the same spectrum has been plotted as a contour plot (*left*) and a pseudocolor plot (*middle*), together with a color calibration bar (*right*). Contour plots allow more detailed examination of the structure of peaks, but estimation of relative intensities is easier with pseudocolor representation.

of peaks. Another advantage is that they can be rapidly manipulated, changing the correspondence between intensity and color, zooming (selecting a portion of the spectrum and magnifying it) and panning (changing the location of the selected portion) to illustrate specific parts of the spectrum in greater detail. Modern graphics displays often have sufficient "video RAM" (special-purpose memory for storing pixel values) to accommodate large portions of the spectrum, allowing manipulations to be carried out on the entire image at video rates of 30 to 100 times a second.

For all their speed, pseudocolor representations have a subtle but significant drawback. This stems from the way in which the human visual system processes color. Smooth variations in color can appear to be discontinuous, under certain circumstances, giving the impression of sharp edges where there are none. The illusory edges that appear to be present in monochromatic images having intensity variations are called Mach bands, and they are a widely studied psychophysical phenomenon. Also, two regions that have the same color can appear to be different colors, and vice versa, depending on the appearance of the surrounding regions; this phenomenon was explored in great detail by Edwin H. Land [60].

A different use of color is for classification, rather than representing a continuous variable. A simple example is to use color in contour plots to designate levels that have opposite sign (Fig. 7.4A). Some colors typically appear more "distant" or "cooler" than others; the "distant" color blue can be used to

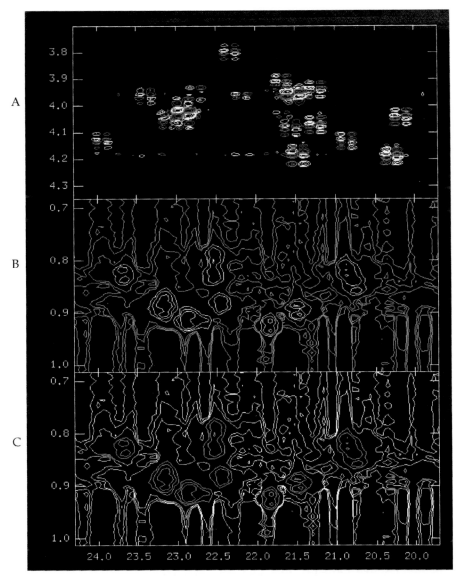

Fig. 7.4. Color serves well for categorization. In (**A**), it is used to distinguish positive and negative contour levels. Careless choices can make it more difficult to discern information: (**B**) and (**C**) differ only in their color assignments, but the peaks are easier to discern in (**B**).

plot contour levels near the noise level while "warmer" colors such as red or orange can be used for higher contour levels. The result is a plot in which peaks clearly stand out from the background, avoiding the visual "clutter" that occurs if the same color is used for all contour levels. Proper choice of colors is important. Although Figure 7.4B and C contain the same information, the use of white for high contour levels makes it easier to discern the peaks in Figure 7.4B than in Figure 7.4C, where blue is used instead.

Some additional factors influence the optimal use of color. One is the distinction between "figure" and "ground": The figure is the information being conveyed and ground is the background on which the information is superimposed. The background is usually black or white. Obviously, white stands out most clearly against a black background, and vice versa. A less obvious effect of the background occurs with depth cueing: Subtle changes in intensity are easier to discern against a black background than against a white background. Another factor to keep in mind when using color is the nature of the display medium. Color computer displays and color printing usually use different systems for representing colors: CRT displays usually use RGB (red, green, and blue) as the colors of the phosphors in the screen, whereas color printing is usually done using CMYK (cyan, magenta, yellow, and black) as the colors of the inks. Colors that appear quite vivid in one representation can appear muddy when displayed using the other representation. For example, bright blue against a black background is easy to discern on a color computer screen, but is difficult to discern when printed, since blue requires a mixture of inks. (To avoid this problem, the "blue" in Fig. 7.4 is actually cyan.) Yet one more consideration for the choice of colors comes from the prevalence of red-green color blindness. In any large audience, there is a significant likelihood that some members will be unable to distinguish green from red. One solution is to avoid using both red and green in the same figure. Alternatively, additional cues can be supplied, such as line thickness or line type (dashed or dotted).

7.1.4. Three-dimensional Spectra

Three-dimensional spectra have three independent dimensions and one dependent dimension; we are running out of display dimensions! However, it is possible to combine the notion of contour plots with perspective views to convey all four dimensions at once. Contours of constant intensity now correspond to three-dimensional surfaces, and they can be rendered with perspective (Fig. 7.5). To avoid overly confusing images, usually only one contour level is drawn. As with perspective views of two-dimensional spectra, perspective views of three-dimensional contours rapidly convey relative positions of features, but they are not useful for determining positions quantitatively. Also, as for two-dimensional perspective views, features in the foreground can obscure features in the background. We usually find contour plots of two-dimensional cross sections of three-dimensional (or higher-dimensional) spectra to be more useful.

7.1.5. The Importance of Being Consistent

As remarkable as it is, the visual system can be overwhelmed quite easily. Truly effective displays avoid unnecessary "chart clutter" and provide consistent visual cues. Multiple representations of a spectrum can be very helpful: For example, combining two-dimensional contour plots together with one-

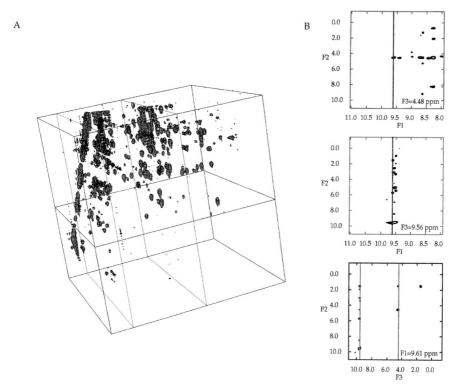

Fig. 7.5. (**A**) Three-dimensional spectra can be displayed by contour plots drawn in perspective. While these are helpful for giving a general overview of the data, contour plots of two-dimensional cross sections (**B**) are better for conveying quantitative information. Color coding can indicate the orientation and relation between the different views, helping the viewer to avoid becoming lost in space.

dimensional views of individual cross sections, or three-dimensional contour plots together with contour plots of two-dimensional cross sections. Figures 7.5 and 7.6 present multiple views of a three-dimensional and a two-dimensional spectrum, respectively. In addition to consistent orientation and scale, color is used to identify cross sections. Multiple cursors (lines that act as two- or three-dimensional rulers), with one or more coordinates matched, can emphasize correlations, even within perspective views.

Consistency is especially important when comparing multiple spectra. Corresponding dimensions should be displayed with the same orientation and scale. Graphics workstations can serve as digital light boxes, capable of such feats as superimposing two contour plots, and interactively alternating colors or intensities at the touch of a key. (It was once common practice to compare contour plots of two-dimensional data by plotting them on separate sheets of translucent paper and then superimposing them on a box or table designed to provide illumination from below.) Alternatively, identical or related views of different spectra can be displayed side by side, using matched and color coded cursors to help convey correlations (Fig. 7.7).

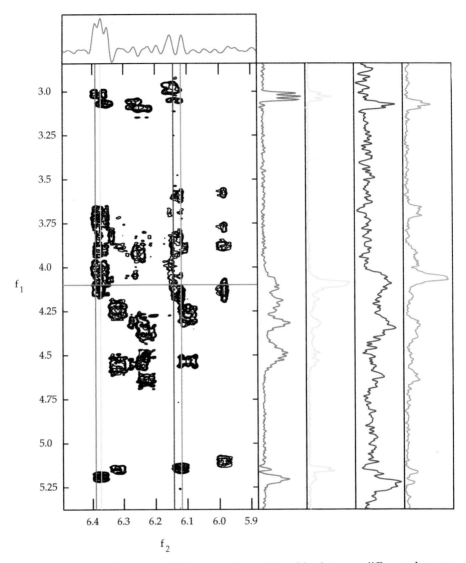

Fig. 7.6. Color coding is useful for expressing relationships between different elements of a display. Here each one-dimensional cross section is drawn in the color of the line indicating the corresponding location in the contour plot. Such visual cues are indispensable for even moderately complicated displays.

Fig. 7.7. Comparison and correlation of multiple spectra require consistent display; in particular, the orientation and scales must match. Spectra can be superimposed (**A**), or displayed side-by-side (**B** and **C**), using matched and color coded cursors.

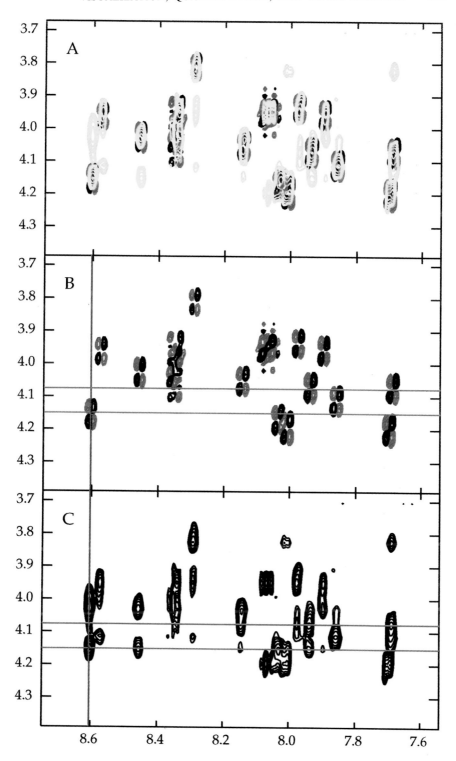

7.2. QUANTIFICATION

Seeing may be believing, but numbers are the distilled essence of any experiment and the foundation for scientific interpretation. In his novel *Arrowsmith* [61], Sinclair Lewis romanticized the power and allure of quantitative research, as opposed to qualitative studies, in a speech by the character Professor Gottlieb:

> Physical chemistry is power, it is exactness, it is life! But organic chemistry—
> that is a trade for pot washers.

(We hasten to add that some of our best friends are organic chemists!) Quantification in NMR, which is the process of obtaining numerical estimates of the characteristics of spectral features, may not be romantic, but it is important.

Parametric methods, such as those based on LP or Bayesian analysis, by their very nature provide quantitative characteristics of peaks. For other spectral estimates, we need separate algorithms for quantifying the location, intensity, and width of peaks.

7.2.1. Peak Positions

The most important characteristic of a spectral peak is its position in the spectrum. Its frequency (or frequencies for multidimensional spectra) contains clues to the chemical nature of the corresponding nucleus and the identity of neighboring nuclei. The position of a peak is usually deemed to be its center of symmetry, to the extent that it has one. Both Lorentzian and Gaussian line shapes have a center of symmetry.

In the absence of extensive overlap, there are several simple methods for estimating the center of a peak. The most common one is to determine the position of maximum intensity; unfortunately, the precision is limited by the digital resolution of the spectrum. Another simple method, that is often capable of precision significantly better than the digital resolution, is to compute the "center of mass" of the peak. In one dimension the center of mass is defined by

$$C_m = \frac{\sum_j f_j (j/N\Delta t)}{\sum_j f_j} \tag{7.2}$$

where the summation is over points that span the peak, from baseline to baseline. Remarkably, the precision of this method improves as the peak gets broader, so long as there is no significant overlap with other peaks. A curved or slanted baseline can bias the result, but is usually not a problem if the baseline level does not change very much over the width of the peak. For determining the *relative* positions of two peaks, the difference between the centers of mass is not affected by a slanted baseline if the slant is the same for both peaks.

Often the relative displacement of two peaks is more important than their absolute frequencies. A somewhat more sophisticated method for determining the relative displacement is to replicate the shape of one peak numerically and then translate and scale the replica to match the other peak. Cubic splines work quite well for replicating peak shapes. (For more about cubic splines, we refer you to *Numerical Recipes in Fortran* [24]. We use the subroutines given there whenever we need to ''spline.'') Suppose one peak spans the frequencies w_j, and another peak the frequencies w_k. Compute a set of spline coefficients that define a smooth curve $C(w)$ (constructed from third-order polynomials) having the same value as the data whenever w is equal to one of the frequencies of the first peak. Since this curve is continuous, it can be used to match the other peak even if the displacement is not an integer multiple of the frequency spacing of the DFT spectrum. The least-squares criterion for the best match is to minimize

$$\sum_k |f_k - aC(w_k + d)|^2 \tag{7.3}$$

where f_k is the intensity of the second peak at frequency w_k, d is the signed relative displacement, and a is a scale factor. If we ignore the imaginary part of $\mathbf{f}$, the scale factor works out to be

$$a = \frac{\sum_k f_k C(w_k + d)}{\sum_k [C(w_k + d)]^2}, \tag{7.4}$$

which is approximately the ratio of the integrals of the two peaks. Since a is determined by Eq. (7.4), minimization of (7.3) requires only a one-dimensional search over d. This method works for arbitrary peak shapes, whether they are symmetric or not; it only requires that the two peaks have approximately the same shape. It can also be used when part of a peak is obscured by an artifact or overlap with another peak, by replicating only the unobscured portion of the peak (restricting the range of j and k appropriately). It is possible to estimate the precision of the result by interchanging the roles of the two peaks and comparing the computed displacements.

The error associated with using the peak maximum, center of mass, and spline shift methods for estimating the relative positions of two peaks is shown in Figure 7.8, based on simulated spectra. Using the peak maximum method, the average error will be comparable to one half the digital resolution. Both the center of mass and spline methods consider all of the points making up the peak to determine its position, not just the maximum, so they are capable of estimating frequencies with a precision far better than the digital resolution. Of course, when peaks span only a few points (or just a single point) of the

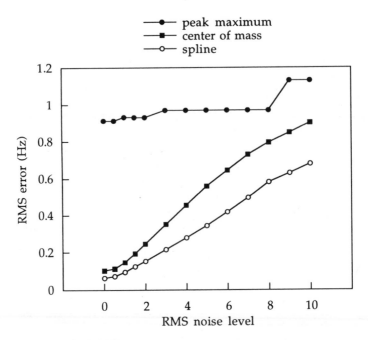

Fig. 7.8. Relative peak displacements can be calculated from the peak maxima, from the centers of mass, or by the method of spline shifts. This graph shows the errors obtained using each method as a function of the noise level in a synthetic spectrum. Exponentially decaying sinusoids with amplitude equal to 10, line width equal to 5 Hz, and frequencies randomly chosen between ±2000 Hz were generated, random noise of varying amplitude was added, and the result was Fourier transformed. The digital frequency resolution of the final spectra was 1.953 Hz. Both the center of mass and spline shift methods are capable of precision substantially exceeding the digital frequency resolution.

spectrum, these methods will not perform much better than using the peak maximum.

All three methods generalize to multidimensional spectra. However, it is simpler to determine the peak positions in different dimensions by using one-dimensional traces through the peaks, parallel to the axis for each dimension.

When overlap is substantial, the best approach is to use modeling, based on known (or presumed) line shapes. A particularly convenient method for fitting Lorentzians to overlapping peaks is to extract a small portion of the spectrum that contains the peaks and compute the inverse DFT, obtaining a mock FID. This mock FID can then be subjected to parametric spectrum analysis, such as LP-SVD, to obtain the peak positions, amplitudes, and widths. For a short mock FID containing just a few peaks, parametric spectrum analysis is relatively efficient. This method is sometimes referred to as ''zooming.''

7.2.2. Peak Integrals

The amplitude of a resonance is proportional to the integral of the corresponding peak in the spectrum (the area under the peak for one-dimensional data, the volume for two-dimensional data, and so on); this is how relative concentrations, relaxation rates, and other rate constants are determined. In nuclear Overhauser spectra, the amplitude is related to the distance between spins, which is the principal basis for NMR as a tool in structural molecular biology.

Precise estimation of peak integrals is complicated by several factors, including the discrete nature of the spectrum, nonzero baselines, overlap, and noise. Many people avoid these difficulties by using crude characterizations of peak volumes such as "strong," "medium," or "weak." It has been argued that such characterizations are sufficient for NMR determinations of biomolecular structure. In other applications, such as the determination of purity or the confirmation of the structure of small organic molecules, quantitative estimates are usually more important.

Integration of functions, also called *quadrature*, is one of the oldest applications of numerical analysis. The literature is full of sophisticated methods, many of which rely on special choices for the points where the function is evaluated (again, we recommend *Numerical Recipes in Fortran* [24] for a detailed and highly readable account). Integration of a function that is defined at uniform intervals is a special case. No survivor of freshman calculus could fail to have been exposed to the methods known as Simpson's rule and the trapezoid rule. Many more sophisticated quadrature schemes exist. Unfortunately, for all their sophistication, these methods don't provide much more accurate estimates of peak integrals, since the results can be no more accurate than the data to be integrated. In NMR, the spectrum is subject to uncertainty as a result of variations in the baseline (or base plane for two-dimensional experiments) and noise. These uncertainties usually dominate the errors inherent in numerical quadrature, and as a result, simple quadrature methods are usually accurate enough. Accuracy can best be improved by reducing overlap, improving S/N, and minimizing artifacts.

In almost all NMR experiments, it is the *relative* areas (or volumes) that are important. The use of relative areas means that the absolute instrument response does not matter, and it has a number of practical implications. One is that the spacing between data points, Δw, is a common factor and can be dropped from all of the quadrature formulae with no loss of generality. The sum rule, trapezoid rule, and Simpson's rule estimates of the area under a peak can then be written

$$area_{\text{sum}} = \sum_{i_{\min} \leq i \leq i_{\max}} f_i \qquad (7.5)$$

$$area_{\text{trapezoid}} = \sum_{i_{\min} \leq i < i_{\max}} \frac{f_i + f_{i+1}}{2} \qquad (7.6)$$

and

$$area_{Simpson} = \tfrac{1}{3} (f_{min} + 4f_{min+1} + 2f_{min+2}$$
$$+ \cdots + 2f_{max-2} + 4f_{max-1} + f_{max}) \qquad (7.7)$$

(Simpson's rule requires an odd number of points.) The differences between these methods lie in the assumptions they make about values of the function between the tabulated values. Simple summation treats the spectrum as having constant value between points. The trapezoid rule uses linear interpolation between points, and Simpson's rule uses parabolic interpolation.

An example will demonstrate the effects of the choice of quadrature scheme and the choice of integration limits. Consider the following estimates of the area of the left-most peak in Figure 7.9A, based on 15 points under the peak (see the close-up in Fig. 7.9B).

Table 7.1

Limits		Area		
		Sum	Trapezoid	Simpson
196	210	22.93	23.00	21.35
197	211	23.06	23.06	24.73
198	212	23.05	23.49	21.67

If the limits used to estimate the integral are shifted one point to the left or right, the Simpson's rule estimate changes dramatically, while the simpler sum and trapezoid estimates are less affected. Evidently Simpson's rule is not well suited to estimating the area of sharp peaks; this can be understood by noting that shifting the integration limits in Eq. (7.7) by one point will change the weight of interior points between 2/3 and 4/3. Similar results are obtained when the integration limits are kept centered at the same location, but expanded or shrunk.

Table 7.2

Limits		Area		
		Sum	Trapezoid	Simpson
196	212	23.46	23.72	21.95
197	211	23.06	23.06	24.73
198	210	22.52	22.78	21.08

For peaks that have similar shapes and line widths, the relative areas can be estimated from *partial* areas, for example the area contained above the half-maximum height. (This is because two peaks with the same shape and line width are related by a vertical scaling—neglecting the effects of discretization.)

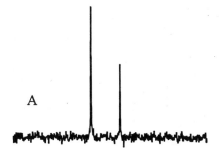

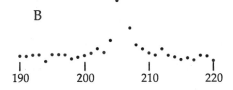

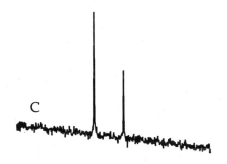

Fig. 7.9. The discrete nature of spectra and artifacts such as sloping baselines can reduce the accuracy of peak area estimates. Parameter estimates for these synthetic spectra, shown in Tables 7.1–7.4, illustrate the effects. (**A**) contains a spectrum with two peaks in a 2:1 ratio. (**C**) is identical to (**A**), except that a linearly sloping base line has been introduced. The close-up of the left-most peak, in (**B**), emphasizes the discreteness of the spectrum.

You can use this fact to minimize the influence of peak overlap on area estimates. One simple method is to fix the limits of integration by choosing a threshold level as a fixed fraction of the peak height: All points above the threshold are used to compute the partial area. This removes the need to explicitly choose upper and lower limits, but it still doesn't eliminate the sensitivity of the estimates to the discrete nature of the spectrum. Consider the following estimates, computed using the trapezoid rule, of the relative areas of the two peaks in Figure 7.9A (the true value is 2.00).

Table 7.3

Threshold Level (%)	Relative Area
50	2.13
25	2.24
10	2.14
5	1.99
3	1.97

If the baseline is truly flat, integrating to lower levels will give a more accurate estimate, as long as overlap between adjacent peaks does not become significant. Conversely, when overlap is more of a problem, the use of higher levels will yield more accurate results. But not if the level is *too* high; area estimates based on only a few points should always be considered suspect.

Which brings us to the problem of estimating areas when the baseline is *not* flat. Figure 7.9C shows the same spectrum as Figure 7.9A, but with a baseline that has an exaggerated slope. The ratios of the areas of the two peaks, using the trapezoid rule, are given in Table 7.4.

Table 7.4

Left Peak Limits		Right Peak Limits		Relative Area
204	208	280	284	1.86
202	210	278	286	1.75
200	212	276	288	1.71
198	214	274	218	1.67
194	218	270	294	1.63

The contribution of the baseline to the integral becomes more significant as the integration limits are expanded. With a steeper slope or wider integration limits, the computed area will become meaningless, since the areas under the baseline overwhelm the areas under the peaks. This exercise shows the importance of correcting the baseline before estimating peak areas.

A very nice method for visually assessing the contribution of the baseline to integral estimates is to plot the indefinite integral as a continuous function, starting from one side of the peak and continuing to the other side (Fig. 7.10). This method dates from the early days of NMR when analog integrators were used to estimate peak areas. If the baseline does not make a significant contribution to the integral, the slope of the integral just prior to and immediately after the peak will be zero; a sloped integral indicates that the baseline is contributing significantly. Analog integrators also allowed a constant slope to be added to the integral, so that if the baseline slope was nearly constant on both sides of the peak, the integral of the baseline could be rendered flat. The corrected integral could be easily estimated as the difference in the integral values on the two sides of the peak. Although analog integrators have given

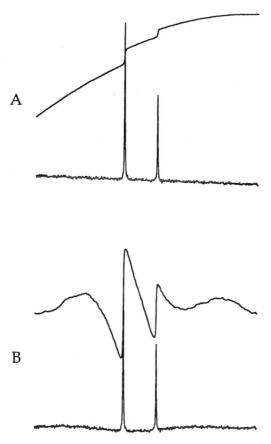

Fig. 7.10. Plotting the indefinite integral of the spectrum as a continuous function is a useful diagnostic for the influence of the baseline slope or curvature on peak area estimates, as well as a simple means to reduce this influence. In (**A**), an estimate of the peak area that does not include a contribution from the baseline slope can be obtained by extrapolating the linear portions of the integral flanking a peak, and determining the vertical offset between the two extrapolations. (**B**) The presence of a curved or nonzero baseline makes area determination more difficult. (The integral is plotted at a lower scale in (**A**) than in (**B**) to keep it from overflowing the bounds of the figure.) In both rows, the integral has been superimposed on the spectrum, to evoke the sense that existed in the days of flatbed plotters and balky liquid ink pens that plotting surface was valuable real estate. Of course, today there is little reason to be so penurious.

way to digital computers and bit-mapped graphics terminals, this scheme is still useful for evaluating and correcting the contribution of the baseline to integral estimates. It has the advantage that the slope correction can be adjusted for each peak separately; the baseline curvature can be complicated, but only has to be approximately constant over the width of a peak for the method to work well. Figure 7.10 shows a modern incarnation of this approach, using the trapezoid rule to compute the continuous integral, rather than an analog integrator.

Numerical quadrature generalizes quite readily to integration in higher dimensions, for example estimating peak volumes in two-dimensional spectra. Each cross section (parallel to the f_2 axis) is integrated just as a regular one-dimensional peak, and the resulting values (parallel to the f_1 axis) are integrated again to yield the peak volume. All of the same caveats apply: Baseline levels (base planes in two-dimensional spectra) and discretization can limit the accuracy of integral estimates. No fundamentally new ideas are required.

7.3. ERROR ANALYSIS

Using quantitative estimates without some notion of the uncertainty in the estimates is a prescription for disaster. It's wise to keep in mind that there are two ways to characterize errors: accuracy and precision. *Accuracy* is a measure of how close an estimate is to the "true" value. *Precision* is loosely used to refer to two distinct, but related concepts. The stated precision of a measurement is the number of significant digits in the numerical value, or equivalently, the log of the relative uncertainty. The repeatability of a measurement refers to the distribution of values obtained if the measurement is made many times. Ideally, the stated precision should agree with the repeatability, which would then be just the precision of the measurement. But sometimes the two can differ—obviously, the number of significant digits in the reported value is entirely up to the discretion of the investigator. Precision is also sometimes expressed as confidence limits, but doing so requires some assumptions about the distribution of errors. The important point to keep in mind is that a highly precise result is not necessarily an accurate one. Oleg Jardetzky has produced a particularly vivid example of the difference between precision and accuracy, and since we know a good idea when we steal one, we include it here—with Oleg's kind permission. The upper panel of Figure 7.11 is an example of a precise and accurate result, while the lower panel represents one that is merely precise.

Error analysis is the study of the uncertainty of quantitative estimates. A large body of formal methods exists for characterizing precision, but accuracy—the "truth"—is harder to pin down. Independent observations that lead to the same conclusion hint at the "truth," but don't necessarily establish it. (Karl Popper's famous edict is that experiments can disprove, but not prove, an hypothesis.) The methods discussed in this chapter can be used to assess both precision and accuracy.

7.3.1. Zero-order Error Analysis: the Laugh Test

A result that leads you to laugh in disbelief has failed the laugh test. The most important question one can ask of a result is: "Is it plausible?" This is a rather

Fig. 7.11. This famous example of the distinction between accuracy and precision is courtesy of Oleg Jardetzky. The archery in the upper panel is both accurate (the arrows strike high-valued regions of the target) and precise (the arrows are tightly clustered around the same position). The aiming in the lower panel is precise, but not accurate.

broad question, so let us try to be more specific. Is it consistent with what we know or believe? If we change the way we process the data, does the result change in an appropriate manner? A result for which the answer to these questions is "No" fails the laugh test; it lacks plausibility since it is inconsistent with the most obvious characteristics of a true result. Many people consider Ockham's razor (the principle that simplest is best) to be an indispensable test of plausibility. Perhaps more important is the principle that extraordinary results demand extraordinary justification.

7.3.2. First-order Error Analysis: Error Propagation

The idea of changing the way the data are processed and seeing if the results are consistent can be formalized and turned into a quantitative assessment of precision. The resulting formalism is called *error propagation*, and with certain simplifying assumptions it takes a particularly elegant form.

Suppose **O** is a set of observations with errors ΔO_i. A result R that is some function f of the observations then has an error ΔR given by

$$\Delta R \approx \sum_i \frac{\partial f(\mathbf{O})}{\partial O_i} \Delta O_i \tag{7.8}$$

(which is just a first-order Taylor expansion), if the errors in the observations are small enough so that the partial derivatives do not vary significantly over the range of the errors. Assuming that the errors ΔO_i are random and independent, the variance in the result [i.e., the squared error, $\text{var}(R) = (\Delta R)^2$] is related to the variance in the observations [$\text{var}(O_i) = (\Delta O_i)^2$] by the formula

$$\text{var}(R) = \sum_i \left(\frac{\partial f}{\partial O_i} \right)^2 \text{var}(O_i) \tag{7.9}$$

since the cross terms vanish. In order to propagate the error using these formulas, we need to know about the function f (in particular, its partial derivatives) in addition to the magnitude of the errors in the observations.

7.3.3. *In Situ* Error Analysis

Sometimes the partial derivatives in the formula for propagation of error can be determined analytically, or they may be known from prior studies. When they are not known, a useful approach is the "injected peak" or *in situ* paradigm. Rather than propagating the error through the steps of data processing, simulated peaks with characteristics close to those of the real ones are added to the raw data *before* any processing steps are applied. Both the real and simulated frequency components are sampled identically and subjected to the same transformations. The characteristics of the simulated signals are known *a priori*, so it is easy to assess the errors in quantifying their parameters. The advantage of this approach is that all the influences of the nonidealities of real data and of the processing steps are included in the error analysis. The presence of nonrandom noise, artifacts, curving baselines, and window functions affects the quantification of real and synthetic peaks alike, so systematic and random errors can both be assessed. However, this approach cannot detect errors in the original measurements; it only uncovers errors arising from processing and quantification.

To perform *in situ* error analysis, the spectrum is first processed (without window functions) to determine the phase corrections in each dimension, to estimate the line widths of the peaks to be analyzed, and to identify suitable frequencies for the injected peaks. Simulated peaks [computed using Eqs. (3.31) and (3.32)] are added to the experimental data, using phases, amplitudes and line widths distributed around those of the peaks to be analyzed. The frequencies of the injected peaks should be chosen carefully to avoid overlap with real peaks. The combined data are processed normally (including appropriate window functions, or MaxEnt, or whatever). The uncertainties in peak positions and amplitudes are determined by comparing the estimated values with the known values for the synthetic peaks.

Figure 7.12 shows a portion of a two-dimensional NOESY spectrum to which simulated peaks have been added; Figure 7.13 compares the estimated and true parameters of the synthetic peaks. (The estimated values were obtained using a "peak picking" routine implemented by Sven Hyberts; it uses the trapezoid rule to estimate volumes above a fractional threshold, here 10% of the peak maximum, and the center of mass to estimate frequencies.) The sim-

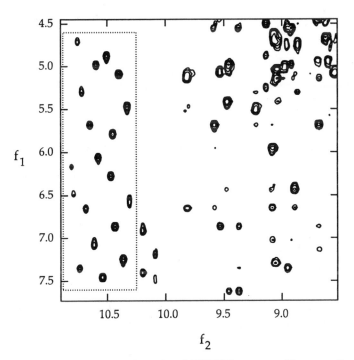

Fig. 7.12. This portion of a two-dimensional NOESY spectrum illustrates the method of *in situ* error analysis. Synthetic peaks that have frequencies lying within the dotted rectangle (an otherwise blank region of the spectrum) were added to the time domain data before any processing steps were performed. Note the asymmetries in the synthetic peaks; these are caused by imperfections in the experimental data. Quantitative estimates of the peak parameters for the synthetic peaks are compared with the known values in Figure 7.13.

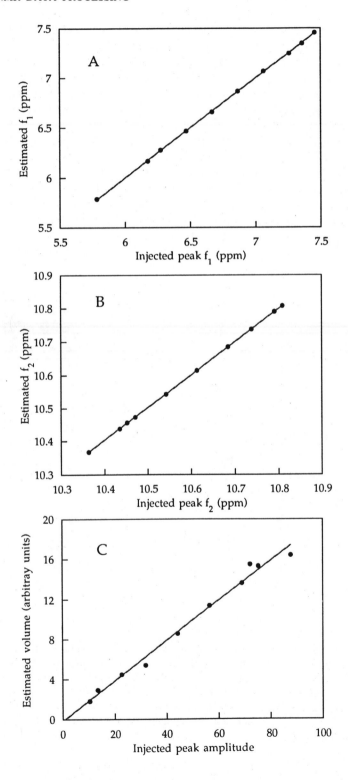

plest measure of the error is the RMS difference between the known parameter values and the parameter estimates: The RMS differences in the chemical shifts are 0.00635 ppm for f_1 and 0.00266 ppm for f_2. In this spectrum the digital resolution is 0.0537 ppm (21.48 Hz) in f_1 and 0.00335 ppm (1.34 Hz) in f_2. For both dimensions the errors are smaller than the digital resolution, but the difference is most pronounced for f_1, which has the lower digital resolution.

When there are systematic errors in quantification, such as those introduced by nonlinear processing methods like MaxEnt, *in situ* error analysis can be used to compensate. With linear processing, it is sufficient to fit a line to the volume estimates for the injected peaks as a function of the injected peak amplitude. The slope of this line is not particularly meaningful, but the intercept should be close to zero; large values indicate a significant base plane contribution. Compensated amplitude estimates are obtained by inverting the equation for the line, and the precision of the estimates can be obtained from the goodness-of-fit of the line to the data. Figure 7.13C shows the line fit to the volume estimates for the injected peaks in Figure 7.12. The intercept is indeed close to zero, and the average error, which is given by the standard deviation of the compensated amplitude estimates divided by the injected amplitudes, is 8%. To compensate for possible nonlinearities, a smooth curve (perhaps a cubic spline, or a more complicated model function if the form of the bias is known) can be fit to the volume estimates for the injected peaks, instead of a straight line. Interpolation using the curve yields compensated amplitude estimates, and, as before, the goodness-of-fit of the compensated estimates to the injected peak amplitudes is a measure of the precision.

The differences between the compensated amplitude estimates derived from Figure 7.13C and the injected peak amplitudes are shown in Figure 7.14. These point-by-point discrepancies are the residual errors, or *residuals*. The residuals of a proper fit to any data should have two properties: the errors should be distributed about a mean value of zero (otherwise it would hardly be the *best* fit), and they should not have a systematic trend (if they do, it suggests that the fitting function is not correct). The residuals plotted in Figure 7.14 have both these properties.

We have pointed out previously that S/N and sensitivity are not the same, and this distinction is particularly important to bear in mind when using nonlinear methods, such as MaxEnt reconstruction. S/N is relatively easy to determine, but it doesn't always give an accurate reflection of the ability to distinguish signals from noise. *In situ* error analysis permits a more useful characterization of sensitivity, especially relative sensitivity (when comparing different processing protocols applied to the same data, for example). The

---

Fig. 7.13. Errors in the quantitative analysis of peaks can be determined by comparing parameter estimates with known values for synthetic peaks added to the experimental data (see Fig. 7.12). (**C**) shows that the peak volumes have been estimated with lower accuracy than the frequencies (**A** and **B**).

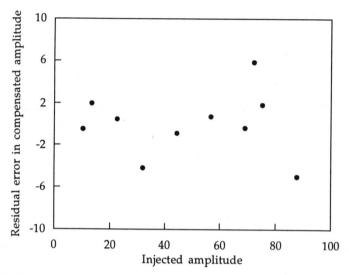

Fig. 7.14. The residual errors of the fit in Figure 7.13C are calculated by inverting the equation of the line to obtain the compensated amplitude estimates: the amplitudes which best fit the measured peak volumes. The residual is the difference between the compensated amplitude estimate and the actual injected peak amplitude, as plotted here. Note that the residuals have an average of zero and show no particular pattern.

amplitudes of the weakest synthetic peaks that can be detected give a better indication of the relative sensitivity of two spectra than their S/Ns.

TO READ FURTHER

A very beautiful book that conveys the power and subtlety of scientific visualization has been written by Tufte [62]. In addition, we recommend the discussion of statistics in *Numerical Recipes in Fortran* [24]. The book by Shoemaker, Garland, and Steinfeld contains a lucid account of elementary error analysis [63].

REFERENCES

[1] R. R. Ernst, "Without computers—No modern NMR," in *Computational Aspects of the Study of Biological Macromolecules by NMR Spectroscopy* (J. C. Hoch et al., eds.), Plenum, New York (1991), pp. 1–26.

[2] D. M. Grant and P. K. Harris (eds.), *Encyclopedia of NMR*, John Wiley & Sons, Chichester (1996).

[3] E. O. Brigham, *The Fast Fourier Transform*, Prentice-Hall, Englewood Cliffs (1974).

[4] J. W. Cooley and J. W. Tukey, *Mathematics of Computation* **19**, 297–301 (1965).

[5] R. R. Ernst and W. A. Anderson, *Review of Scientific Instruments* **37**, 93–102 (1966).

[6] R. N. Bracewell, *The Fourier Transform and Its Applications*, 2nd ed., McGraw-Hill, New York (1978).

[7] R. R. Ernst, *Adv. Magn. Reson.* **2**, 1 (1966).

[8] A. G. Redfield and S. D. Kunz, *J. Magn. Reson.* **19**, 250 (1975).

[9] R. A. Beckman and E. R. P. Zuiderweg, *J. Magn. Reson. A* **113**, 223 (1995).

[10] D. Marion, M. Ikura, and A. Bax, *J. Magn. Reson.* **84**, 425 (1989).

[11] D. J. States, R. A. Haberkorn, and D. J. Ruben, *J. Magn. Reson.* **48**, 286 (1982).

[12] D. Marion and A. Bax, *J. Magn. Reson.* **80**, 528 (1988).

[13] J. Cavanagh and M. Rance, *J. Magn. Reson.* **88**, 72 (1990).

[14] E. Fukushima, *Practical NMR, A Nuts and Bolts Approach*, Addison-Wesley, New York (1981).

[15] D. Shaw, *Fourier Transform NMR Spectroscopy*, Elsevier Science, New York (1976).

[16] T. Farrar and E. Becker, *Pulse and Fourier Transform NMR*, Academic Press, New York (1971).

[17] A. Bax, *Two-Dimensional Nuclear Magnetic Resonance in Liquids*, Delft University Press, Dordrecht (1982).

[18] K. Wüthrich, *NMR of Proteins and Amino Acids*, John Wiley & Sons, New York (1986).

[19] R. Freeman, *Handbook of Nuclear Magnetic Resonance*, John Wiley & Sons, New York (1988).

[20] A. G. Marshall and F. R. Verdun, *Fourier Transforms in NMR, Optical, and Mass Spectrometry: A User's Handbook*, Elsevier Science, Amsterdam (1990).

[21] H. Gesmar, J. J. Led, and F. Abildgaard, *Prog. NMR Spec.* **22**, 255 (1990).

[22] R. Kumaresan, *I.E.E.E. Trans.* **ASSP-31**, 217 (1983).

[23] G. Zhu and A. Bax, *J. Magn. Reson.* **100**, 202 (1992).

[24] W. H. Press, B. P. Flannery, S. A. Teukolsky, and W. T. Vetterling, *Numerical Recipes in Fortran*, 2nd ed., Cambridge University Press (1992).

[25] C. F. Tirendi and J. F. Martin, *J. Magn. Reson.* **85**, 162 (1989).

[26] H. Barkhuijsen, R. de Beer, and D. van Ormondt, *J. Magn. Reson.* **73**, 553 (1987).

[27] J. A. Cadzow, *IEEE Trans.* **ASSP-36**, 49 (1988).

[28] H. Chen, S. van Huffel, C. Decanniere, and P. van Hecke, *J. Magn. Reson. A* **109**, 246 (1994).

[29] J. P. Burg, "A new analysis technique for time series data," in *Modern Spectrum Analysis* (D. J. Childers, ed.), I.E.E.E. Press, New York (1978), pp. 42–48.

[30] J. Tang and J. R. Norris, *J. Chem. Phys.* **84**, 5210 (1986).

[31] G. Zhu and A. Bax, *J. Magn. Reson.* **90**, 405 (1990).

[32] P. Koehl, C. Ling, and J. F. Lefèvre, *J. Magn. Reson. A* **109**, 32 (1994).

[33] B. Svejgaard, *Bit* **7**, 240 (1967).

[34] R. de Beer and D. van Ormondt, *NMR Basic Prin. Prog.* **26**, 201 (1992).

[35] Plato, *Republic*, Book VII. There are many translations of this work, which is thought to have been written around 380 BC. One translation is by G. M. A. Grube, with revisions by C. D. C. Reeve, Hackett Publishing, Indianapolis (1992).

[36] C. E. Shannon, *Bell Syst. Tech. J.* **27**, 379 (1948).

[37] D. L. Donoho, I. M. Johnstone, J. C. Hoch, and A. S. Stern, *J. R. Stat. Soc. B* **54**, 41 (1992).

[38] G. J. Daniell and P. J. Hore, *J. Magn. Reson.* **84**, 515 (1989).

[39] J. Skilling and R. Bryan, *Mon. Not. R. Astron. Soc.* **211**, 111 (1984).

[40] J. A. Jones and P. J. Hore, *J. Magn. Reson.* **92**, 276 (1991).

[41] P. Schmieder, A. S. Stern, G. Wagner, and J. C. Hoch, *J. Biol. NMR* **4**, 483 (1994).

[42] J. A. Jones and P. J. Hore, *J. Magn. Reson.* **92**, 363 (1991).

[43] P. F. Fougère (ed.), *Maximum Entropy and Bayesian Methods*, Kluwer Academic, Dordrecht (1990).

[44] B. Buck and V. A. Macaulay (eds.), *Maximum Entropy in Action: A Collection of Expository Essays*, Oxford University Press, New York (1991).

[45] C. R. Smith and W. T. Grandy, Jr. (eds.), *Maximum-Entropy and Bayesian Methods in Inverse Problems*, D. Reidel, Dordrecht (1985).

[46] B. F. Logan, Ph.D. Thesis, Columbia University (1965).

[47] A. Papoulis, *I.E.E.E. Trans. Circ. Syst.* **CAS-22**, 735 (1975).

[48] P. A. Jansson, R. H. Hunt, and E. K. Plyler, *J. Opt. Soc. Am.* **60**, 596 (1970); see also F. Ni and H. A. Scheraga, *J. Magn. Reson.* **82**, 413 (1989).

[49] S. J. Davies, C. Bauer, P. J. Hore, and R. Freeman, *J. Magn. Reson.* **76**, 476 (1988).

[50] G. Strang, *SIAM Review* **31**, 614 (1989).

[51] D. L. Donoho, Technical Report No. 437, Department of Statistics, Stanford University (1993).

[52] G. L. Bretthorst, *J. Magn. Reson.* **88**, 533, 552, and 571 (1990).

[53] S. Sibisi, "Quantified MaxEnt: An NMR application," in *Maximum Entropy and Bayesian Methods* (P.F. Fougère, ed.) Kluwer Academic, Dordrecht (1990), pp. 351–358.

[54] R. A. Chylla and J. L. Markley, *J. Biomol. NMR* **3**, 515 (1993).

[55] C. R. Lindberg, Ph.D. Thesis, University of California at San Diego (1986).

[56] G. Strang, *American Scientist* **82**, 250 (1994).

[57] A. Papoulis, *Signal Analysis*, McGraw-Hill, New York (1977).

[58] A. V. Oppenheim and R. W. Schafer, *Digital Signal Processing*, Prentice-Hall, Englewood Cliffs (1975).

[59] J. D. Foley, A. van Dam, S. K. Feiner, and J. F. Hughes, *Computer Graphics: Principles and Practice*, 2nd ed., Addison-Wesley, Reading (1990).

[60] E. H. Land, *Scientific American* **237**, 108 (1977).

[61] S. Lewis, *Arrowsmith*, Harcourt Brace Jovanovich, New York (1953).

[62] E. R. Tufte, *Visual Display of Quantitative Information*, Graphic Press, Cheshire (1983).

[63] D. P. Shoemaker, C. W. Garland, and J. I. Steinfeld, *Experiments in Physical Chemistry*, 3rd ed., McGraw-Hill, New York (1974).

INDEX